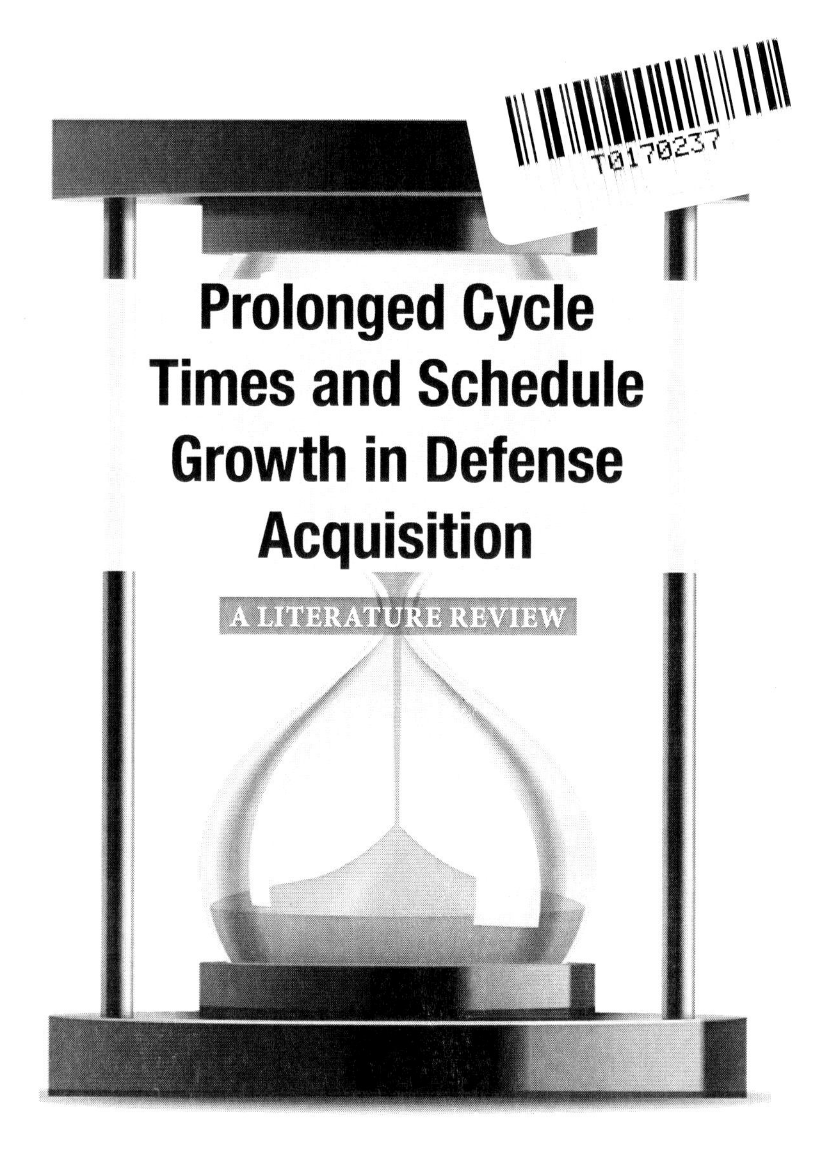

Prolonged Cycle Times and Schedule Growth in Defense Acquisition

A Literature Review

Jessie Riposo, Megan McKernan, Chelsea Kaihoi Duran

Prepared for the Office of the Secretary of Defense

The research described in this report was prepared for the Office of the Secretary of Defense (OSD). The research was conducted within the RAND National Defense Research Institute, a federally funded research and development center sponsored by OSD, the Joint Staff, the Unified Combatant Commands, the Navy, the Marine Corps, the defense agencies, and the defense Intelligence Community under Contract W91WAW-12-C-0030.

Library of Congress Cataloging-in-Publication Data is available for this publication.

ISBN: 978-0-8330-8515-3

The RAND Corporation is a nonprofit institution that helps improve policy and decisionmaking through research and analysis. RAND's publications do not necessarily reflect the opinions of its research clients and sponsors.

Cover design by Dori Walker / Hour glass image: Ion Popa/Fotolia

RAND OFFICES

SANTA MONICA, CA • WASHINGTON, DC

PITTSBURGH, PA • NEW ORLEANS, LA • JACKSON, MS • BOSTON, MA

CAMBRIDGE, UK • BRUSSELS, BE

www.rand.org

Preface

Lengthy or delayed acquisitions may translate into a critical delay of necessary capabilities to the warfighter and additional costs to the government. While there has been extensive research into weapon systems cost growth, the research to objectively determine the primary causes of longer cycle time and schedule growth is less comprehensive. Schedule management is challenging because schedule is intrinsically tied to many other aspects of acquisition.

Although large numbers of studies have focused on or provided insights into schedule-related issues, these issues are less well understood than other aspects of acquisition. This report summarizes the assertions and conclusions from such sources. We do not analyze these assertions and conclusions; rather, we provide an accounting and summary of the range of claims in the literature at various periods. Moreover, while current conditions may differ from history, this report and the sources described herein provide a starting point for examining the schedule-related aspects of acquisition.

This report should be of interest to government acquisition professionals, oversight organizations, and, especially, the analytic community as a starting point for further research and analysis.

This research was sponsored by the Director, Acquisition Resources and Analysis, in the Office of the Under Secretary of Defense for Acquisition, Technology, and Logistics, U.S. Department of Defense. This research was conducted within the Acquisition and Technology Policy Center of the RAND National Defense Research Institute, a federally funded research and development center sponsored by the

Office of the Secretary of Defense, the Joint Staff, the Unified Combatant Commands, the Navy, the Marine Corps, the defense agencies, and the defense Intelligence Community. For more information on the Acquisition and Technology Policy Center, see http://www.rand.org/nsrd/ndri/centers/atp.html or contact the director (contact information is provided on the web page).

Contents

Tables

Summary

This report summarizes a selection of the acquisition literature from the 1960s to the present on potential sources of program schedule cycle time and growth, as well as potential opportunities for improvement. It presents the range of possible causes of schedule-related problems and various recommendations cited for improving schedules by various authors and organizations. This report does not provide critical analysis or an assessment of the strengths or weaknesses of the claims made in the literature. Rather, it provides a starting point for further research or consideration by government acquisition professionals, oversight organizations, and the analytic community.

Potential Reasons for Longer Cycle Times or Schedule Delays

In documentation accompanying the release of the Better Buying Power 2.0 program, Frank Kendall, Under Secretary of Defense for Acquisition, Technology, and Logistics, stressed the need to "reduce cycle times while ensuring sound investment decisions." He added that, on average, programs are taking about one year longer to complete their development contracts than they did before 1980; the root causes of longer program cycle times are not obvious, and the data include wide variations (Kendall, 2013; OUSD[AT&L], 2013). In defense acquisition, *cycle time* is often defined as program initiation[1] to initial operating capability (IOC), though it can also be defined in other ways, depending on the specific focuses of analyses of acquisition program life cycles (e.g., time from Milestone A to Milestone B or time from Milestone B to Milestone C). Like the cycle times of acquisition programs, schedule growth—the extension to the planned schedule—can also be measured.

One of the goals in managing an acquisition program is to create a realistic schedule based on technological maturity, system complexity, and anticipated budget. We identified the following reasons for schedule delays in the literature:

- The reason asserted most often is the **difficulty of managing technical risk** (e.g., program complexity, immature technology, and unanticipated technical issues).
- The second most common reason is **initial assumptions or expectations that were difficult to fulfill** (e.g., schedule estimates, risk control, requirements, and performance assumptions).
- Another common assertion is that **funding instability** complicates management and can directly stretch production schedules.

Table S.1 presents the full list of general causal categories cited in the literature that can affect longer cycle times and schedule delays, along with attendant detailed characterizations as presented by various studies.[2] Some of these processes and activities occur outside of the control of program management, while others fall under the control of program management. Note that the listed reasons are not necessarily internally consistent (i.e., some studies assert that some of these possibilities are not major concerns).

---

[1] Formal program initiation is normally at Milestone B, but some may consider earlier program initiation points, such as Milestone A or even the prior Materiel Development Decision (MDD).

[2] Note that some of these may appear to be similar or overlapping. We sought to preserve the focus and characterization of the original studies in formulating this table.

Table S.1
Reasons Cited in the Literature for Prolonged Schedules and Schedule Slippage

Area	Possible Reason
Requirements development, generation, and management	Infeasible or unrealistic requirements
	Unstable requirements (e.g., engineering requirements, readiness requirements, reliability and support requirements)
	Inefficiencies in the process (e.g., serial nature of process and requirements evolution)
Managing technical risk	Excessive technical, manufacturing, or integration risk (general) or program complexity
	Unanticipated design, engineering, manufacturing, technical difficulty, or technology integration issues
	Overly optimistic assumptions/expectations (technical risks, performance goals, system requirements, or design maturity)
	Immature technology
	Concurrency in complicated programs
	Prototyping
	Deficient test planning or testing inefficiencies
	Inadequate funds for testing
Resource allocation	Funding instability or budget cuts
Defense acquisition management	Lack of focus on schedule or inadequate schedule management (e.g., underutilization of integrated master schedule)
	Overly optimistic assumptions/expectations in general, including insufficient contingency funds in program budgets
	Overly optimistic assumptions/expectations in cost and schedule estimates
	Personnel issues
	Competition
	Use of undefinitized contract actions
	Contractor performance and inadequate incentives
	Inadequate tailoring of the acquisition process
Other	Delays in obtaining necessary data

Potential Ways to Remedy Schedule Challenges

Improving a schedule is about meeting the mission and threat in a more timely fashion. In the case of a low-risk program employing commercially available technologies, schedule improvement may mean a tailored reporting and oversight process that allows a program to waive some portions of the acquisition process that may be more appropriate for a new-build, higher-risk program. In the case of a high-risk program requiring the development and integration of new technologies, schedule improvement may mean better risk management and more compromises. These insights may lead to decisions to reduce the program scope, or even to cancel the program, to free up resources for more executable programs. All of these decisions are essentially about reducing and managing risk. Because of heightened interest in reducing cycle times, this report summarizes published recommendations for shortening timelines (ideally without adverse consequences). In the literature, the most commonly cited recommendations for reducing cycle time and controlling schedule growth are **strategies that manage or reduce technical risk**. Some of those recommendations include

- using incremental fielding or evolutionary acquisition (EA) strategies
- developing derivative products (rather than brand-new designs)
- using mature or proven technology (i.e., commercial, off-the-shelf components).

Other recommendations include maintaining stable funding and using atypical contracting vehicles. No strategy is appropriate in every case, so careful judgment and balancing are required. Table S.2 summarizes the strategies suggested in the literature.

Table S.2
Possible Ways to Improve Schedules in the Acquisition Literature

Area	Possible Ways to Improve Schedules
Requirements development, generation, and management	Stable and realistic initial requirements, especially at the engineering level
	Better collaboration between the program management and end-user communities (with proper management)
	Proper management of flexible requirements
Managing technical risk in development and production	Use of mature/demonstrated technology to ensure a high level of maturity before production
	Use of incremental fielding or EA strategies and the development of derivative products (rather than brand-new designs)
	Employment of "agile" methods that can easily adapt to changes in software development
	Prototyping
	Concurrency in programs with low technical risk
	Use of commercially derived items
	Use of the commercial practice of freezing the design before the production contract award
	Use of the commercial practice of reducing the design's complexity
Resource allocation	Stable funding
	Adequate test funds (hardware, modeling and simulation)
Acquisition management: internal to the program	Bypassing competition during production (including employing multiyear or sole-source procurement strategies in the production phase)
	Preplanned product improvement
	Acquisition of the same number of units but in larger, more economical quantities in the production phase
	Emphasis and adherence to schedule as a program priority
	Development and maintenance of a comprehensive and realistic master schedule

Table S.2—continued

Area	Possible Ways to Improve Schedules
Acquisition management: internal to the program (continued)	Use of contracting vehicles to expedite contracting process (e.g., existing contracts, undefinitized contracts in low-rate initial production, sole-source contracts)
	Operational testing and evaluation results available before production startup
	Use of modeling and simulation to reduce the risk and duration of live tests
	Involvement of the test community in all program phases
	Use of integrated product teams
	Improved program stability in general, including funding and requirements
	Realistic schedule estimates
Acquisition management: external to the program	Senior leadership support
	Program identified as a priority

Acknowledgments

We thank the researchers who have addressed schedule issues in the research that informs this literature review. We also thank the many RAND researchers who provided guidance and support to this project: Irv Blickstein, Jeffrey Drezner, Charles Nemfakos, John Graser, Christopher Pernin, Bruce Held, William Shelton, Robert Leonard, Akilah Wallace, Michael McMahon, Mark Arena, John Schank, and James Kallimani. We also would like to thank Kate Giglio, Lauren Skrabala, and Maria Falvo, who assisted in structuring, editing, and preparing this document for publication. In addition, we are especially grateful to Philip Antón and Mark Lorell, who provided detailed reviews that greatly improved this report.

Finally, we would like to thank Michael Rich, president and chief executive officer of RAND, for his support, along with the director of the RAND Acquisition and Technology Policy Center, Cynthia Cook, and the associate director, Paul DeLuca, for their insightful comments on this research.

Abbreviations

ACAT	Acquisition Category
AMRAAM	Advanced Medium-Range Air-to-Air Missile
APB	Acquisition Program Baseline
BBP	Better Buying Power
CPD	capability production document
DAU	Defense Acquisition University
DoD	U.S. Department of Defense
DSB	Defense Science Board
EA	evolutionary acquisition
FLTSATCOM	Fleet Satellite Communication System
FRP	full-rate production
GAO	U.S. General Accounting Office (prior to July 7, 2004) U.S. Government Accountability Office (as of July 7, 2004)
IDIQ	indefinite delivery/indefinite quantity
IED	improvised explosive device

IMS	integrated master schedule
IOC	initial operating capability
IT	information technology
JSF	Joint Strike Fighter
JUON	joint urgent operational need
LANTIRN	Low-Altitude Navigation and Targeting Infrared for Night
LRIP	low-rate initial production
MDAP	major defense acquisition program
MDD	Materiel Development Decision
MRAP	Mine-Resistant Ambush-Protected
OSD	Office of the Secretary of Defense
OUSD(AT&L)	Office of the Under Secretary of Defense (Acquisition, Technology, and Logistics)
OT&E	operational testing and evaluation
PARCA	Performance Assessments and Root Cause Analyses
SBIRS	Space-Based Infrared System
TRL	Technology Readiness Level
UCA	undefinitized contract action
UON	urgent operational need
USD(AT&L)	Under Secretary of Defense for Acquisition, Technology, and Logistics
WSARA	Weapon Systems Acquisition Reform Act

CHAPTER ONE

Background and Motivation

Lengthy acquisitions or unexpected schedule slips in military system acquisition may translate to a delay in delivering capabilities to the warfighter or additional unexpected costs to the government. The U.S. Department of Defense's (DoD's) recent Better Buying Power (BBP) initiative is aimed at "obtaining greater efficiency and productivity in defense spending" (Carter, 2010a, p. 1), including a focus on shortening program cycle times as part of its guidance for improving the acquisition process:

- **BBP 1.0:** "Set shorter program timelines and manage to them," stated then–Under Secretary of Defense for Acquisition, Technology, and Logistics (USD[AT&L]) Ashton Carter (Carter, 2010a, pp. 4–5).
- **BBP 2.0, introduction to the acquisition workforce:** "Reducing cycle times while ensuring sound investment decisions" is the initiative's goal, according to USD(AT&L) Frank Kendall (Kendall, 2012a, p. 2).

Kendall has further honed in on schedule cycle time or lengthy acquisitions, recently worrying that it is "taking much too long" to field systems (Parrish, 2012). In the BBP 2.0 implementation directive, released April 24, 2013, Kendall states,

> At this time, the data is not clear as to the effect of the acquisition process itself on cycle time. Most decision support activities overlap with program progress, so in general the decision making pro-

> cesses add overhead more than direct schedule slips. Nevertheless, reducing the burden of this overhead is a worthwhile goal in its own right. On average programs are taking about 1 year longer to complete development than they did 20 years ago, but the root causes of longer program cycle times are not obvious and the data includes wide variations. Time is money, and slowness in acquiring major systems does mean greater expense and fewer capabilities in the field. There have been attempts to use arbitrary cycle times to constrain programs; however, these constraints have often been unrealistic and done more harm than good by leading to high risk schedules and acquisition approaches. During 2013, we will conduct additional analysis of the time it takes from conception to introduce a product to the field. Under BBP 2.0, we will focus on reducing the decision making cycle time and overhead costs while those studies are being conducted. (Kendall, 2013, p. 16)

Kendall's statement regarding increased cycle times is supported by the annual report *Performance of the Defense Acquisition System,* released by his office in June 2013 (OUSD[AT&L], 2013). That analysis of a set of contract schedule and cost growth data from 1970 to 2011 concluded the following:

- All other things equal, development cycle time on contracts after 1980 took an average of 0.9 years longer than contracts before 1980—an increase of about one-sixth over the base of 5.2 years.
- Every 10-percentage-point increase in work-content cost growth generally added 0.066 years, and every 10-percentage-point increase in cost-over-target generally added 0.16 years.
- Also, contracts with undefinitized contract actions (UCAs) were about 0.3 years longer generally.
- Contracts for space systems were an additional 1.7 years longer, whereas contracts for aircraft were 2.5 years longer. No other commodity types had significantly longer cycle times.

- All development contract cycle times increased significantly after 1980. (OUSD[AT&L], 2013, pp. 55, 58)

Literature Review Goals, Methodology, and Limitations

Cycle times and schedule growth—along with cost growth—have always been concerns for DoD acquisition programs; however, the literature has historically focused more directly on cost growth. We conducted this literature review to capture available insights on the experiences of a variety of programs and experts to illuminate *schedule* issues and potential solutions. The intent of this report is to summarize the acquisition literature describing sources of excessive cycle time and schedule growth, as well as opportunities for reducing the time it takes to deliver systems from the perspective of shortening individual program schedules. This report does not provide critical analysis of the assertions of the various organizations and authors. Rather, it is intended to serve as a starting point for further research or consideration by government acquisition professionals, oversight organizations, and the analytic community.

Our methodology included a broad search of government, academic, and nonprofit analytic sources as far back as the 1960s.[1] We also consulted subject-matter experts at RAND to help focus the review. We looked for instances in which authors or organizations provided reasons for schedule growth or increased cycle times. We also searched for recommendations for remedying these problems. Finally, we sorted and presented the reasons and recommendations by topic and source. (The reasons and recommendations most commonly addressed in the literature are discussed in greater detail in Chapters Two and Three.) This report identifies factors both internal and external to programs that influence schedule plans and schedule outcomes.

---

[1] Future research should include information on this topic from congressional testimony and Selected Acquisition Reports, which we did not examine because of time constraints. Both of these sources, and possibly some additional sources in the academic and policy literature, would strengthen the results of this literature review.

The Importance of Schedule

According to the Defense Acquisition University's (DAU's) *Scheduling Guide for Program Managers* (2001, p. 1), a schedule, "in its simplest form . . . is a listing of activities and events organized by time." But, in its more complex form, a schedule is defined as follows:

> [T]he process examines all program activities and their relationships to each other in terms of realistic constraints of time, funds, and people, i.e., resources. In program management practice, the schedule is a powerful planning, control, and communications tool that, when properly executed, supports time and cost estimates, opens communications among personnel involved in program activities, and establishes a commitment to program activities. (DAU, 2001, p. 1)

When looking at changes in acquisition schedules, cycle times of programs should be considered alongside schedule growth. Cycle time can be defined by how long it takes a program to get from one part of the defense acquisition process to another. This could be from program initiation[2] to initial operating capability (IOC), "Milestone A to Milestone B," or other periods (e.g., Milestone B to Milestone C).

In addition, the Office of the Under Secretary of Defense for Acquisition, Technology, and Logistics (OUSD[AT&L]) stresses that acquisition is also about managing risk and uncertainty:

> Acquisition is about risk management—not certainties. Especially for major weapons systems acquisitions (which almost always involve research and development), uncertainties imply cost, schedule, and performance risks relative to early estimates. These risks diminish as we move from research to development through production to sustainment, but their realization may result in cost and schedule growth. These risks also require use of different management tools (such as the right contract types

---

[2] Formal program initiation is normally at Milestone B, but some may consider earlier program initiation points such as Milestone A or even the prior Materiel Development Decision (MDD).

> and incentives) at different stages to mitigate risks and motivate industry to achieve the lowest possible total price to the government. We must monitor and explain risks, but it is important to remember that developing technologically superior military capability is not a risk-free endeavor. (OUSD[AT&L], 2013, p. 109)

What Lengthens Acquisition Schedules?

The most pressing concerns related to poor schedule outcomes seem rooted in two areas: **maintaining a technological edge** and **excessive costs**. In a letter preceding the 2012 Strategic Guidance for 21st Century Defense (DoD, 2012a), then–Secretary of Defense Leon Panetta described tomorrow's military as "a Joint Force for the future that will be smaller and leaner, but will be agile, flexible, ready, and technologically advanced. It will have cutting edge capabilities, exploiting our technological, joint, and networked advantage." Fulfilling this vision requires the acquisition process to support the timely delivery of systems and capabilities. Longer development, production, and fielding times increase the risk that the technology may not be adequate to address current threats or may become obsolete in the face of emerging threats shortly after deployment. On the other hand, artificially accelerated programs risk schedule and cost growth as reality sets in or with the deployment of immature technologies or flawed weapon systems. This may, in turn, result in the delivery of reduced capabilities and an increased need for modifications and maintenance in the future, both of which incur costs and limit the availability of new systems. Finally, the delayed development, production, and fielding of new systems can ultimately result in smaller or older fleets, which may be inadequate to face adversaries (see DoD, 2012a).

In the current fiscal environment, the importance of controlling costs is self-evident. In 2010, Ashton Carter, then USD(AT&L), described the relationship between cost and schedule across program portfolios in his BBP 1.0 guidance to DoD acquisition professionals using common perceptions that have been since supported by significantly more data and analysis (also see OUSD[AT&L], 2013):

> The leisurely 10–15 year schedule of even the simplest and least ambitious Department programs not only delays the delivery of needed capability to the warfighter, but directly affects program cost. As all programs compete for funding, the usual result is that a program settles into a level-of-effort times the length of the program. Thus a one-year extension of a program set to complete in 10 years can be expected to result in 10 percent growth in cost as the team working on the project is kept on another year. Yet managers who run into a problem in program execution generally cannot easily compromise requirements and face an uphill battle to obtain more than their budgeted level of funding. The frequent result is a stretch in the schedule. (Carter, 2010a, pp. 4–5)

Relationships between cost and schedule have also been widely discussed in the literature. For example, numerous studies have reported that programs with longer durations tend to have greater cost growth, possibly because longer programs are exposed to more opportunities for changing requirements and other time-related costs (see, for example, Arena, Leonard, et al., 2006; Drezner et al., 1993; Arena, Younossi, et al., 2006; OUSD[AT&L], 2013; and Drezner and Smith, 1990).[3] However, this relationship between cost and schedule is not as simple as it may appear. Schedule slippage and cost growth are often thought to go hand in hand. OUSD(AT&L) recently examined some of the complexities in the requirements, technology, cost, and schedule growth relationship:

> Contextually, we note that the time required to acquire next-generation capabilities is often longer than the strategic threat and technology cycles these capabilities are meant to address. Performance (good or bad) in planned defense acquisition is intertwined with cost and schedule implications from unplanned responses to these external demands. This is not an excuse for cost and schedule growth, but an observation from first principles that changing threats and needs can add costs and delays relative to original

[3] Drezner and Smith (1990, p. 43) state, "An overly lengthy program can cost more than a shorter program, all other things being equal, in large part because of the inflation and overhead allocation, and the opportunity to change requirements."

> baselines as ongoing acquisitions are adjusted. (OUSD[AT&L], 2013, p. 109)

Finally, it is important to realize that these relationships are rarely seen as bidirectional (i.e., shortening schedules probably will not alleviate a tight budget). While longer programs tend to incur cost growth, a program with an aggressive schedule will tend to incur both more work annually and added rush costs, thus yielding higher annual and total costs.

Can Acquisition Schedules Be Shortened?

Long schedules may be unavoidable in high-risk technology programs because the technology will take some time to mature, and because added funding has limitations in shortening development. Also, schedule slippage can occur in any program, whether it involves high-risk technology or not (e.g., when the planned schedule is unrealistically short for a given task, or when management issues arise). Determining an appropriate or optimal schedule for an acquisition program, and adhering to it, requires a delicate balancing act that should include taking into account the program's technological maturity, complexity, and budget. As Pernin and colleagues assert in their report, *Lessons from the Army's Future Combat Systems Program*,

> Any acquisition program faces the dual risks that the future capabilities envisioned today may not meet the actual operational needs of tomorrow, and that technological progress simply may not occur as quickly as anticipated. The longer the timeline, the more uncertain the future becomes, which amplifies the first risk; but with more time for technology to mature, in some ways, a longer timeline also dampens the second risk. (Pernin et al., 2012, p. 52)

A longer timeline, as discussed above, can allow more time for the technology to mature in an acquisition program, but it also allows the program more time to adjust to changes that it may encounter. In some cases, programs may experience negative consequences when accelerating acquisition. For example, Hanks et al. (2005) note that speeding up the contracting process may lead to vagueness in the way contracts are written, potentially causing problems with contractors who could use the ambiguity to their advantage rather than the government's. In the Advanced Medium-Range Air-to-Air Missile (AMRAAM) major defense acquisition program (MDAP), shortened schedules caused design tasks to slip from demonstration and validation to full-scale development to meet milestone dates (Mayer, 1993). Others have noted that the emphasis on streamlining and scheduling has also been problematic because there are not enough opportunities for trade-offs among cost, schedule, and performance (see, for example, Hanks et al., 2005).

More rapid acquisition, in addition to not always being desirable, is also not always plausible. Technological maturity—specifically, when the technology used in an acquisition program has not reached the planned level of maturity—is one of the most widely cited causes of lengthy acquisition schedules, but it is not the only roadblock to accelerating the acquisition process. Aggressively accelerated schedules require access to higher yearly funding and other resources, which are competed for across portfolios of programs and, thus, are allocated based on the priority of the program within the portfolio. Typically, only high-priority programs enjoy the resourcing necessary to accommodate an optimal schedule (see, for example, Younossi et al., 2007). Additionally, the processes that regulate system acquisition were put in place to ensure that programs are managed effectively and in such a way that the government receives high-quality products. As discussed later, there are areas where these processes could be improved, but as a rule, these processes cannot—and should not—be ignored.

The amount of time it takes to deliver a system to the end user is a function of what is being procured and how. DoD procures new systems and upgrades to existing systems. New systems may be revolutionary, evolutions of existing platforms, or commercial products or derivatives. The levels of design, development, and production involved, and therefore of schedule risk and mitigation, vary for each type of new system or upgrade.

Three interdependent systems govern how defense materiel is procured: the Joint Capabilities Integration and Development System

(JCIDS) Requirements System (see CJCSI 3170.01H, 2012); the resource allocation system known as the Planning, Programming, Budgeting, and Execution System (see DoDD 7045.14, 2013); and the Defense Acquisition System (see DoDD 5000.01, 2003; DoDI 5000.02, 2008; Interim DoDI 5000.02, 2013). The operation of each of these systems affects how quickly materiel needs can be satisfied.

What Does It Mean to "Improve" a Schedule?

If a lengthy acquisition system is unacceptable, but universally accelerating program schedules could lead to undesirable results and is not always feasible, what exactly does it mean to "improve" a schedule? Improving a schedule is about meeting the mission and threat in a more timely fashion. If this proves too difficult, then alternatives need to be explored. In the case of a low-risk program employing commercially available technologies, schedule improvement may mean a tailored reporting and oversight process that allows a program to waive some portions of the acquisition process that may be more appropriate for a new-build, higher-risk program. In the case of a high-risk program requiring the development and integration of new technologies, schedule improvement may mean better risk management and more compromises. These insights may lead to decisions to reduce the program scope, or even cancel the program, to free up resources for more executable programs. All of these decisions are essentially about reducing and managing risk. Because of heightened interest in reducing cycle times, this report summarizes published recommendations for shortening timelines (ideally without adverse consequences).

DoD schedule estimation and adherence are affected by many factors, including some that are internal and external to program office and government control. Making the evaluation of schedule even more complex are the interrelationships among these factors that affect schedule. For example, the resource allocation system and acquisition system are not mutually exclusive. As a result, schedule activities cannot be evaluated in isolation.

Schedule Improvement Is an Ongoing Challenge and Goal

Schedule improvement is not a new objective for DoD. Ward and Quaid (2006, p. 14) stated that the "need to decrease the technology development timeline predates the Revolutionary War." In 1986, the Packard Commission Report declared that "an unreasonably long acquisition cycle . . . is a central problem from which most other acquisition problems stem" (Packard et al., 1986, p. 47). According to the commission, "In frustration, many have come to accept the ten-to-fifteen-year acquisition cycle as normal, or even inevitable. We believe that it is possible to cut this cycle in half" (Packard et al., 1986, p. 52).

The Government Performance and Results Act of 1993 (Pub. L. 103-62) required government agencies to develop a strategic plan for program activities and to report annually on performance goals. One of DoD's early performance goals was associated with improving acquisition, including reducing cycle times. DoD's performance report for fiscal year (FY) 2000 claimed that the department had accomplished its goal of reducing the amount of time it takes an MDAP to get from program initiation[4] to IOC from 132 months to under 99 months (DoD, 2001, p. 49). The report cites advanced concept technology demonstrations, integrated product teams, and "commercially derived items" as means to achieve this goal. However, the DoD Inspector General reported, "The database used to calculate MDAP acquisition cycle time for inclusion in the FY 2000 Annual Report . . . was not accurate or complete."[5] We were unable to identify a revised report with a new estimate for MDAP cycle times. In DoD's FY 2013 budget proposal, the cycle time goal was as follows: "5.3.3-2E: Beginning in FY 2011, the DoD will not increase by more than five percent from the Acquisition Program Baseline (APB) cycle time for Major Defense

---

[4] Program start was defined as "MS [milestone] I, MS II or MS III," depending on the first major milestone for each individual acquisition program.

[5] The report continues, "Of the 48 MDAPs reviewed, data for 28 programs was incorrect. We also identified three programs that were not included in the database. As a result of our findings, USD(AT&L) has contracted for the complete verification and reconciliation of any omissions and inconsistencies in the database. As of December 2001, USD(AT&L) estimated that it will complete the verification and reconciliation of the database by February 2002" (Office of the Inspector General, U.S. Department of Defense, 2001, p. i).

Acquisition Programs (MDAPs) starting in FY 2002 and after." DoD reported that it met its goal because the actual cycle time increase in FY 2011 was 4.5 percent (DoD, 2012b).

The U.S. Government Accountability Office (GAO) has used multiple methods over the past 40 years to examine cycle times and schedule delays, but there has not been a consistent reporting track or data set for the results. This makes it difficult to draw broad conclusions about the department's progress. One GAO report stated that "five major studies, which cover the period from 1970 to 1986, show that the problems being experienced today in the weapons acquisition process are similar to those of the past" (GAO, 1988, opening letter). Less than ten years ago, GAO discussed persistent problems that plagued weapons systems, concluding that "defense acquisition programs in the past 3 decades continued to routinely experience cost overruns, schedule slips, and performance shortfalls" (GAO, 2006, p. 4).

The above discussion on schedule increases and defense acquisition programs illustrates two points:

1. There is no consensus in the literature on whether or not DoD has improved its scheduling efforts over time.
2. There is widespread agreement that technology risk and management issues are the most important causes of schedule slippage.

Organization of This Report

Chapter Two describes the major themes in the literature with respect to sources of schedule growth. Chapter Three discusses the major themes in the literature with respect to schedule growth mitigation and schedule improvement. Chapter Four presents some conclusions based on this literature review. Finally, a case study of the Mine-Resistant Ambush-Protected (MRAP) Vehicle acquisition program is included in the appendix. The MRAP acquisition program, which prioritized schedule and performance, was provided with significant resources and senior leadership support to meet urgent needs in the field.

CHAPTER TWO

Sources of Schedule Growth

In this chapter, we look at the case studies and expert opinions in the literature regarding the causes of increased cycle time or schedule growth in acquisition programs. We first offer a broad overview of the problem and then focus on some of the individual points covered by the literature. The literature describing sources of schedule growth tends to focus on negative, rather than positive, program examples. Thus, many recommendations for schedule improvement focus on avoiding pitfalls rather than ways to achieve shorter cycle times. These pitfalls are discussed in this chapter, while cited recommendations are discussed in Chapter Three.

Reasons for Schedule Growth

Recent work by OUSD(AT&L)'s Performance Assessments and Root Cause Analyses (PARCA) office—supported by research by the Institute for Defense Analyses and the RAND Corporation—focused on uncovering the root causes of Nunn-McCurdy breaches.[1] As a consequence, PARCA has identified an array of primary reasons for cost growth. While a majority of the findings and recommendations in these analyses center around cost growth, the studies also address schedule issues, such as the causes and consequences of schedule slip,

[1] See Bliss, 2012a, 2012b, and 2013; Blickstein et al., 2011; Blickstein et al., 2012a; and Blickstein et al., 2012b.

and successful schedule management techniques. Note that more proximal causes resulting from root causes are not cited in their analyses.

In ten of the 18 root-cause analyses conducted on major acquisition programs with critical cost growth, "poor management performance" was the predominant root cause. This includes poor management of systems engineering, contractual incentives, risk, and situational awareness (OUSD[AT&L], 2013, p. 34). Other root causes included unrealistic baseline costs and schedule estimates, as well as changes in procurement quantity. OUSD(AT&L)'s study also noted that the following causes were sometimes found to be the root cause of cost growth in the 18 programs examined by PARCA: immature technology; excessive manufacturing or integration risk; unrealistic performance expectations; and unanticipated design, engineering, manufacturing, or technology issues (OUSD[AT&L], 2013, p. 34). Most notably, "funding inadequacy or instability" was not a root cause of critical growth in the 18 programs; however, it has been frequently asserted in other literature as a cause of schedule growth in programs. This difference may be due to the strict criteria used by PARCA to define root causes.

There is some overlap in the literature that touches upon the reasons for schedule slippage or growth in acquisition programs and PARCA's root causes of cost growth; however, we discovered many additional reasons for schedule slippage, asserted by various authors and organizations. As stated previously, we did not assess whether these authors or organizations conducted sufficient analysis to support their conclusions; rather, we have tried to create a compilation of what is available in the literature.

Table 2.1 lists the various reasons for longer cycle times and schedule growth, along with the studies that cited them.

It should be noted that rarely, if ever, do problems occur in isolation. Schedules are typically subject to many factors that may not rise to the level of a root cause of cost or schedule growth but could influence the schedule of a program nonetheless. These factors include unrealistic performance expectations; unrealistic baseline estimates of costs and schedules; immature technologies or excessive manufacturing or integration risk; unanticipated design, engineering, manufacturing, or

Table 2.1
Possible Reasons Cited in the Literature for Prolonged Schedules and Schedule Slippage

Area	Possible Reason	Analysis
Requirements development, generation, and management	Infeasible or unrealistic requirements	GAO, 2012c; GAO, 2011a; Bodilly, 1993; Pernin et al., 2012; Comptroller General of the United States, 1979; Decker et al., 2011
	Unstable requirements (e.g., key performance requirements, readiness requirements, reliability and support requirements)	Arena, Birkler, et al., 2005; GAO, 1986c; GAO, 2011a; OUSD(AT&L), 2013, pp. 33–35
	Inefficiencies in requirements process (e.g., serial nature of process and requirements evolution)	Decker et al., 2011; Comptroller General of the United States, 1971; Pernin et al., 2012
Managing technical risk	Excessive technical, manufacturing, or integration risk (general) or program complexity	Blickstein et al., 2011; Blickstein et al., 2012a; Blickstein et al., 2013; GAO, 1979, pp. 2, 10; Tyson et al., 1991; Drezner and Smith, 1990, p. 46; B. Fox et al., 2004; OUSD(AT&L), 2013, pp. 33–35
	Unanticipated design, engineering, manufacturing, technical difficulty, or technology integration issues	Blickstein et al., 2011; Blickstein et al., 2012a; Blickstein et al., 2013; Drezner and Smith, 1990, pp. vii, 44; GAO, 1986b, p. 5; Cashman, 1995, pp. viii, 62; GAO, 1986c, p. 42; OUSD(AT&L), 2013, pp. 33–35
	Overly optimistic assumptions/ expectations (technical risks, performance goals, system requirements, or design maturity)	Blickstein et al., 2011; Blickstein et al., 2012a; Blickstein et al., 2013; Glennan et al., 1993, p. xi; Schinasi, 2008, p. 2; GAO, 1991, p. 4; GAO, 2012b; OUSD(AT&L), 2013, pp. 33–35
	Immature technology	Blickstein et al., 2011; Blickstein et al. 2012a; Blickstein et al., 2013; OUSD(AT&L), 2013, pp. 33–35
	Concurrency in complicated programs	GAO, 2012a, p. 10; Comptroller General of the United States, 1979, p. 10; GAO, 2012b; Younossi et al., 2005; Kendall, 2012b

Table 2.1—continued

Area	Possible Reason	Analysis
Managing technical risk (continued)	Prototyping	Tyson et al., 1991; Drezner and Smith, 1990; Kendall, 2012b
	Deficient test planning or testing inefficiencies	Comptroller General of the United States, 1972, p. 38; B. Fox et al., 2004, pp. xxii, 108
	Inadequate funds for testing	GAO, 2011b, p. 20
Resource allocation	Funding instability or budget cuts	Drezner and Smith, 1990, p. vii; GAO, 1986c, p. 42; J. R. Fox, 2011, p. 98; GAO, 1986b, p. 5; Kassing et al., 2007, p. 8; GAO, 1991, p. 4; Glennan et al., 1993, p. xi; OUSD(AT&L), 2013, pp. 33–35
Defense acquisition management	Lack of focus on schedule or inadequate schedule management (e.g., underutilization of an integrated master schedule [IMS])	Anderson and Upton, 2012, p. 36; Drezner and Smith, 1990; Farr, Johnson, and Birmingham, 2005
	Overly optimistic assumptions/ expectations in general, including insufficient contingency funds in program budgets	GAO, 1986c, p. 42; Younossi et al., 2005; OUSD(AT&L), 2013, pp. 33–35
	Overly optimistic assumptions/ expectations in cost and schedule estimates	Blickstein et al., 2011; Blickstein et al., 2012a; Blickstein et al., 2013; Glennan et al., 1993, p. xi; GAO, 2012b; Mayer, 1993; Lorell and Graser, 2001; Younossi et al., 2008; GAO, 1986a, pp. 6–7; GAO, 1987, p. 3; GAO, 1991, p. 4; Comptroller General of the United States, 1971, p. 21; OUSD(AT&L), 2013, pp. 33–35
	Personnel issues	Blickstein et al., 2011; Blickstein et al., 2012a; Blickstein et al., 2013; Cashman, 1995, p. viii; Bodilly, 1993
	Competition	Tyson et al., 1989, p. VII-7; Birkler et al., 2001, p. 29; Drezner and Smith, 1990; Gailey, 2002; Reig, 1995
	Use of UCAs used during development	OUSD(AT&L), 2013

Table 2.1—continued

Area	Possible Reason	Analysis
Defense acquisition management (continued)	Contractor performance and inadequate incentives	Cashman, 1995, p. viii
	Inadequate tailoring of the acquisition process	Drezner et al., 2011; Decker et al., 2011
Other	Delays in obtaining necessary data	Cashman, 1995, p. viii

technology integration issues arising during program implementation; and poor performance by government or contractor personnel.[2] These results highlight the sentiment expressed in our many discussions with RAND analysts: Schedule management is difficult because the schedule is intrinsically tied to all other aspects of acquisition.

In addition, GAO has found that schedule slippage can occur in all portions of the acquisition process. In one report, GAO analyzed acquisition programs from the 1970s through 1984. It found that "about 30 percent of the total schedule slippages experienced by the 1970's systems occurred during the 1980s," meaning that the systems were further along in their acquisition life cycles and still experienced schedule slippage (GAO, 1986b, p. 5). Perhaps unsurprisingly, a more recent GAO report stated that "studies of more than 700 defense programs have determined there is limited opportunity for a program to get back on schedule once they are more than 15 to 20 percent complete" (GAO, 2013, p. 7).

Internal and External Activities

The literature also identifies issues and activities that are internal and external to programs or, in other words, within or outside of program management control. Activities and decisions internal to programs include contracting strategies, cost and schedule estimation, and personnel issues.

---

[2] For more examples of root cause analyses, see Blickstein et al., 2011, and Blickstein et al., 2012a.

Influential factors external to programs include stability in funding and requirements, policy changes, and contractor performance. The literature identifies potential sources of schedule growth that fall into several broad categories: technical issues; requirements development, generation, and management; resource allocation; and program management. Prominent themes in the literature include the following:

- difficulty managing technical risk (e.g., program complexity, immature technology, unanticipated technical issues)
- initial assumptions or expectations that are difficult to fulfill (e.g., cost and schedule estimates, risk, requirements, and performance assumptions)
- funding instability.

The following discussion of the literature on causes of schedule growth and cycle time increases starts with requirements development, generation, and management.

Requirements Development, Generation, and Management

Requirements development is a major DoD process whereby capability gaps and desired future capabilities are identified and validated. The function and intent of this process has many implications for a program's schedule. There were two ways identified in the literature in which requirements affect schedule: (1) overly demanding requirements at the beginning of a program and (2) requirements changes throughout a program. Requirements changes can be positive when they involve descoping as a recognition of overly ambitious initial requirements, or they can be negative when they add on and change consistently, as in "requirements creep." GAO (2012c) reported that infeasible, unstable, or overly ambitious requirements can lead to rework and, thus, cost and schedule growth. Infeasible or overly aggressive requirements are also closely linked to program risk. However, it is not always easy to identify whether the changes to requirements or overly ambitious requirements are the cause or effect of schedule growth and other problems. A 2011 GAO report put forward some assertions regarding the complexity of the relationship between changing requirements and cost and schedule growth:

> While changing requirements creates instability and, therefore, can adversely affect program outcomes, it is also possible that some programs experiencing poor outcomes may be decreasing program requirements in an effort to prevent further cost growth. . . . [P]rograms with changes to performance requirements experienced roughly four times more growth in research and development costs and three to five times greater schedule delays compared to programs with unchanged requirements. Similarly, programs with increases to key system attributes—lower level, but still crucial requirements of the system—experienced greater, albeit less pronounced, cost growth and schedule delays than other programs. (GAO, 2011a)

Infeasible or unstable requirements (e.g., key performance parameters, readiness requirements, reliability and support requirements) might be caused by a variety of factors. A report reviewing 22 Acquisition Category (ACAT) I programs terminated since the end of the Cold War identified changes in requirements stemming from changes in leadership, the reprioritization or restructuring of programs (which enabled technology and requirements creep), optimistic Technology Readiness Level (TRL) assessments, and optimistic technology integration and manufacturing readiness assessments (Decker et al., 2011). The same report identified the requirements process itself as a source of longer-than-necessary schedules, noting "an average of 15 months to staff a requirements document for ACAT I programs. The corresponding time to staff an ACAT II program is 18 months, and it is 22 months for an ACAT III program" (Decker et al., 2011, p. 35). In an evaluation of shipbuilding programs, Arena et al. (2005) identified changing requirements in the form of change orders and late production definition to be predominant sources of schedule slip.

Other reported program experiences with regard to requirements and schedule are summarized below. These cases demonstrate challenges in the initial determination of acceptable levels of technical risk and technological maturity. They show that program management cannot always address such risks without exceeding initial cost or schedule estimates:

- The Air Force's Low-Altitude Navigation and Targeting Infrared for Night (LANTIRN) program experienced schedule delays because of ambitious requirements, extensive concurrency, and an inexperienced system program office (Bodilly, 1993).
- Process inefficiencies in the Engineering Change Proposal process and a user community resistant to changing overly ambitious requirements (i.e., the program's framing assumptions) led to significant schedule issues in the Future Combat Systems program (Pernin et al., 2012).
- Two communication satellites—Fleet Satellite Communication System (FLTSATCOM) and the third generation of the Defense Satellite Communication System—faced developmental and technical problems caused by design sophistication, which eventually led to high costs and schedule delays. Additionally, "[t]he stringency of the Navy's and the Air Force's communications requirements for the FLTSATCOM satellites caused technical difficulties in the development program. These difficulties caused cost overruns and schedule delays" (Comptroller General of the United States, 1979, pp. 2–3).

Resource Allocation

Resource allocation is the DoD process that allocates funds to (in part) procure and sustain the broad array of materiel and equipment identified and validated by the requirements process (DoDD 7045.14, 2013). While a description of this process falls outside the scope of this report, the function and intent of the process has many implications for program schedules. Most notably, various reports have asserted that the instability of program funding or budget cuts may lengthen schedules.

The resource allocation process, like the requirements generation process, is lengthy. The Future Years Defense Program is typically formulated every other year and covers a period of six years. The services begin working on the program three years before the first-year funds are appropriated (DoDD 7045.14, 2013). The underlying assumption is that the services can identify what they will need in three years and that those funding requirements will remain stable. McCaffrey and Jones (2005) reported that it is unlikely that the costs of weapon systems estimated three years prior would reflect actual costs. While changes can be made to the Future Years Defense Program, this normally occurs only twice a year—in August or September and in January of the following calendar year (Fast, 2010). The length of this process and limitations on changes make it difficult to adjust program funding for "fact-of-life" changes in a timely manner, requiring funding and program requirements to be forecast accurately.

In addition to the challenges associated with ensuring appropriate funds, programs can face funding instability stemming from multiple sources (e.g., congressionally mandated budget cuts, reprioritizations within service portfolios, comptroller reallocations of unobligated program funds). Drezner and Smith (1990) found that factors external to the program (including funding stability) can have a profound effect on program length. In two case studies of Army programs conducted by Kassing et al. (2007), funding instability came from two sources: events that occurred outside the control of Army leaders and ambitious technical goals set by the Army. In their case study on the Javelin program, "the program approved for development of the Javelin missile system in 1989 was recognized as ambitious at the time. Technical problems followed, and the development schedule had to be extended, resulting in what was high development funding instability by our measure. In addition, before the Javelin could move into production, the Cold War ended, Army forces were cut, and the Javelin procurement objectives were cut nearly in half. These 'fact of life' changes led to high procurement funding instability" (Kassing et al., 2007, p. xvii). However, in this same analysis, the authors found only a small but positive statistical association between total funding instability and schedule slippage, pointing out that funding instability can be either a cause or effect of other problems (Kassing et al., 2007, p. 86).

In addition, Ronald J. Fox has asserted that funding levels and annual budget fluctuations have caused schedule and other problems, particularly in the Cold War era, when "changing funding levels prompted by annual budget fluctuations often led to inefficient production rates and schedule slippages in key weapons programs contracted out to industry" (R. J. Fox, 2011, p. 98).

Technical Risk

According to a GAO report published in 1979, excessive technical risk has been "probably the single most significant factor leading to weapon failures, cost growth and overrun, production interruption or shutdown, production inefficiency, and schedule slippages" (GAO, 1979, p. 10). One source of technical risk is the use of immature technology. Excessive concurrency can increase the cost of fixing problems incurred by prematurely entering into production or the next program phase before the technology and design have fully matured. Prototyping can mature the technology, and early testing can identify problems when they are easier to correct. These management considerations have all been found to carry schedule implications. However, defense acquisition is about pushing the state of the art, so some of these risks will always be present in programs.

Because many aspects of the acquisition process exist primarily to manage technical risk, technical issues are intrinsically tied to several other schedule-influencing factors. A 2005 literature review published in the *Defense Acquisition Review Journal* examined a dozen studies describing sources of schedule slippage, finding the most commonly cited source of schedule slippage to be technical in nature (Monaco and White, 2005). Competition, prototyping, and contract type were the other most prevalent topics discussed with regard to schedule. In a review of ten acquisition programs, Drezner and Smith (1990, p. 44) identified technical difficulties as among the factors with the most profound impact on program length. In an effort to identify sources of schedule problems across large U.S. Air Force system development efforts, Cashman (1995) cited technical problems, delays at the subcontractor level, delays in getting necessary data, manufacturing problems, and staffing problems as the five most common sources of schedule growth. However, he clarified that the most common sources are not necessarily the most problematic in terms of the amount of dollars and time associated with the slip, meaning that it is useful to account for relative importance. He found technical problems and subcontractor delays to be associated with the greatest dollar values, while problems with subcontractor performance and manufacturing had the largest negative impact in terms of time. Technical issues are logically tied to the amount of technical risk that the program can handle, which is influenced by several program characteristics, including the feasibility and stringency of the specific technical and design solution chosen to meet the requirement. Most other program characteristics that influence technical risk fall within the purview of the acquisition management system. For example, sometimes the component technologies may be fairly mature, but the integration of the technologies may be extremely challenging, as is the case in the Space-Based Infrared System (SBIRS) satellite program.

Technology Complexity and Maturity

One of the major sources of technical risk is the use of complicated and immature technologies. While many argue that program durations are getting longer (OUSD[AT&L], 2013), it is often pointed out that program complexity is also increasing, and this increasing complexity may be at fault for longer program timelines (Drezner, 2009, p. 32). Bernard Fox and colleagues suggest a relationship between weapon system complexity and schedule duration: "As military systems have become more complex, testing has become more time consuming and costly" (B. Fox et al., 2004, abstract).

GAO suggests that a majority of programs are employing technology that is *nearing* maturity, but not fully mature, prior to entering system development (GAO, 2012a). Specifically, in an annual review of a sample of 62 programs in 2007, GAO found that only 16 percent had achieved technology maturity at Milestone B. It also found that at the start of production (Milestone C), only 67 percent of the programs sampled had achieved technology maturity (GAO, 2007).

Several analyses performed by Blickstein and colleagues (2011, 2012a, 2012b, and 2013) have identified immature technology as a significant (but not the only) factor in program problems. Program examples include the DDG-1000 destroyer, the Joint Strike Fighter (JSF), the Wideband Global Satellite Communication System, Excalibur, and Apache Block III.

The importance of technological maturity in timely delivery has also been underscored by commercial experience, an example being Boeing's recent experience developing the 787 Dreamliner. The Dreamliner was designed to improve fuel efficiency and ensure a smoother, quieter, and more comfortable ride for passengers. These improvements required a substantial increase in the amount of composite materials relative to older airframes, as well as new manufacturing techniques. At the same time, in an effort to cut costs, Boeing outsourced much of its manufacturing and supply. The resulting effects on development and production schedules have been less than favorable: The program has seen several substantial delays—nearly three years in total—caused by supply shortages, production delays, and testing problems. In 2010, after a fire grounded a test fleet, Boeing's chief executive officer, Jim McNerney, said, "In retrospect, our 787 game plan may have been overly ambitious, incorporating too many firsts all at once—in the application of new technologies, in revolutionary design-and-build processes, and in increased global sourcing of engineering and manufacturing content" (Peterson, 2011).

Concurrency

Concurrency is a strategy in which development and production activities partially overlap in time, rather than being performed sequentially, with the intent to compress program timelines and (ideally) reduce costs. Some modest amount of concurrency is usually prudent, but how soon concurrency should be initiated to balance risk continues to be a point of debate and management attention:

> Concurrency is broadly defined as the overlap between technology development and product development or between product development and production. While some concurrency is understandable, committing to product development before requirements are understood and technologies mature or committing to production and fielding before development is complete is a high-risk strategy that often results in performance shortfalls, expected cost increases, schedule delays, and test problems. It can also create pressure to keep producing to avoid work stoppages. (GAO, 2012b)

Such overlaps can sometimes reduce the overall time required for development and production, but they also carry risks: When design and development problems surface after production has begun, units already produced need to undergo costly retrofits (if they are even possible), adding cost and time (GAO, 1979, p. 10). Overlaps can also cause programs to produce units that are not cost-effective to retrofit and are either inferior or not operational. The role of concurrency in schedule performance is intricately tied to the level of technical risk remaining in a program's design: When design risks are low, concurrency can shorten schedules for delivering operational units at either reduced or modest cost; when technical risks are high and borne out, concurrency can lead to numerous problems, including cost and schedule growth.

The F-35 JSF program is a recent example of a program that had a significant level of concurrency that resulted in problems (Kendall, 2012b; Kendall et al., 2012). The F-22 program is another example in which concurrency reportedly contributed to schedule growth. Several F-22 design challenges—related to the advanced technology being employed but underestimated or unaccounted for in original program plans—caused cost growth and delays. These difficulties were substantially compounded by concurrency in development and integration, which likely led to further delays (Younossi et al., 2005). The GAO also reported that, because of high levels of concurrency, a design problem in the Missile Defense Agency's Ground-Based Midcourse Defense program was discovered after production was under way. The result, according to GAO (2012b), was significant cost and schedule growth, as well as potential retrofits to fielded equipment. GAO has also warned that several current programs—JSF, SBIRS High, and Apache Block IIIA Remanufacture—are at increased manufacturing risk because of their concurrent development and production strategies (GAO, 2012a, p. 10). OUSD(AT&L) also reported that "first principles indicate that concurrent production when designs are unstable can impose added retrofit costs for early production products. Further analysis of past performance will provide an objective foundation for informing future policies and acquisition decisions" (OUSD[AT&L], 2013, p. 109).

Prototyping and Testing

Prototyping has been identified as a potential means for reducing technical and manufacturing risks (see, for example, Tyson et al., 1991). Prototyping can mature the technology, and testing can identify problems early and determine the remaining level of risk. Unfortunately, the implications for program schedule resulting from prototyping efforts are mixed. Drezner and Smith (1990) reported that prototyping lengthens schedules, yet other studies have identified prototyping activity that reduced schedules (see, for example, Tyson et al., 1991), while still others found no such relationship (see, for example, Nelson and Trageser, 1987; Tyson et al., 1989; Harmon et al., 1989). The counterfactual is difficult to assess, however: It is unclear whether the technical risks resolved in prototyping would have led to greater program length than the additional time it took to complete the prototyping activities. It also depends on what is prototyped. For example, the competitive prototyping in the JSF program tested only some characteristics of the fighter airframe and propulsion designs. Risk was probably reduced in these areas, but not in many others. Further, prototyping is not likely to significantly reduce the system integration risk, since full integration amounts to full-scale development.

In a review of Air Force programs, Bernard Fox et al. (2004) suggest that the increasing complexity of military systems has led to longer and more costly testing programs. The authors recommend more testing, which should be accounted for in the original schedule. This may increase cycle times but might not necessarily increase schedule growth, if the original scheduling is accurate. The authors also suggest that the relationship between schedule duration and testing is nuanced because "most T&E expenditures occur in the later stages of development, when cost overruns and schedule slips from other activities may have become more apparent. As a result, there is often considerable pressure to expedite and/or reduce T&E activities to recoup some of the other overruns." The report also describes how "government test range personnel were not as focused on controlling the costs and schedule of the test program as other members of the test team were" (B. Fox et al., 2004, p. xxii). Some individuals interviewed suggested that some test procedures could be changed to "improve the pace and efficiency of the typical test program," suggesting inefficiencies in the testing process (B. Fox et al., 2004, p. xxiii).

The Office of the Comptroller General (the predecessor to GAO) has also asserted that some weapon systems can have schedule growth caused by "deficient test planning." It reported that some test schedules assumed that only minimal problems would surface during testing, schedules lacked contingencies, and planned testing environments were inadequate for proving operational utility because this type of testing was completed too late in the process (Comptroller General of the United States, 1972, p. 38). In a more recent study, the GAO reported that inadequate funding of developmental testing was a source of cost and schedule growth, and that limiting the amount of testing performed increases program risk and could "result in an extension of a program's test schedule" (GAO, 2011b, p. 20).

Defense Acquisition Management: Practices, Policies, and Procedures

According to DoD Directive 5000.01 (2003), "The Defense Acquisition System exists to manage the nation's investments in technologies, programs, and product support necessary to achieve the National Security Strategy and support the United States Armed Forces," with a primary objective of "acquiring quality products that satisfy user needs with measurable improvements to mission capability and operational support, in a timely manner, and at a fair and reasonable price." Managing these investments involves several challenges, including developing and adhering to realistic schedule estimates, establishing an acquisition approach that incentivizes contractors and is appropriately tailored to the program's needs, and managing the risks inherent in technology development. As discussed below, these challenges, if not handled appropriately, can cause numerous problems, including schedule growth.

Program Management: Schedule Estimation and Management

Historically, schedule estimation and management have not received the same level of attention as cost estimation and management within DoD. Anecdotally, many have noted a lack of priority schedule adherence receives in program management (see, for example, Anderson and Upton, 2012, p. 36); others have blamed this lack of emphasis on schedule adherence on deteriorated program control competency[3] and a lack of accountability for timelines (for example, Drezner and Smith, 1990; Farr, Johnson, and Birmingham, 2005). According to Michael Sullivan of GAO,

> Most program managers seem focused on controlling costs and delivering a quality product. The third leg of the acquisition stool 'program schedule' is perceived to be less important and seems to be a resource that can be slipped to accommodate unstable funding or technical difficulties when they are encountered. Given that most major defense program schedules span years or even decades, schedule slips are less likely given their importance. (Sullivan, 2012, p. 23)

According to Farr, Johnson, and Birmingham (2005, p. 237), "The lack of schedule focus creates a culture that has historically called for a total system solution despite notorious requirements creep and a never-ending component to sub system-to-system test." Also, if achieving requirements is the highest priority, then cost growth and schedule

---

[3] Program control competency includes the ability to use and manage tools, which helps with the development and management of schedule.

slippage can result when encountering technical difficulties and funding shortfalls.

Perhaps because of this lack of focus on schedule or the difficulties inherent in schedule estimating, robust schedule-estimation methodologies have not been as extensively developed as cost-estimation methodologies. A 2012 survey of current program managers revealed that a vast majority believed that an integrated, up-to-date schedule is a critical component of successful program management, but fewer than 50 percent were confident that their schedules were accurately resource-loaded. Only 51 percent believed that all government and contractor workload was included in their schedule (Sullivan, 2012, p. 23).

In addition to potential inadequacies in schedule estimating and a lack of focus on schedule, the literature reports cases in which program schedules have suffered negative effects from excessive optimism. For example, OUSD(AT&L)'s *Performance of the Defense Acquisition System: 2013 Annual Report* contains a summary of root-cause analyses of critical cost growth in 18 programs: "Baseline cost and schedule estimates were unrealistic in just over one-fourth of the cases. The primary underlying reason was invalid framing assumptions. . . . Framing assumptions are any explicit or implicit assumptions central in shaping cost, schedule, and/or technical performance expectations" (OUSD[AT&L], 2013, p. 34). In an earlier report, GAO found "that overly optimistic assumptions about technical risks were common factors in [cost and schedule] overruns" (GAO, 1991, p. 4). Another study identified ambitious schedules, budgets without contingency funds, and a reluctance to adjust ambitious performance goals as factors that may influence schedule (Glennan et al., 1993). Ambitious schedules that do not account for risks can set up a program to not meet its original schedule. Reports cite cases in which such problems have occurred as a result of program advocacy, bureaucratic pressure, and operational urgency:

- Mayer (1993) reported that the AMRAAM program vastly oversold itself in terms of cost and schedule. Early cost estimates may have been made for advocacy reasons, and the schedule was compressed (from 90 to 70 months) in response to congressional pressure. This compressed schedule caused design tasks to slip from the demonstration and validation phase of the program to full-scale development. Eventually, a gap between what was advertised and what was achieved emerged due to technical problems.

- Lorell and Graser (2001) found that an overly aggressive schedule was a likely cause of cost growth and schedule restructuring for the SBIRS program. Despite stable requirements for SBIRS High, Younossi et al. (2008) found that most of the program's cost growth was due to inappropriate cost and schedule estimates made by the contractor and accepted by government.
- Blickstein et al. (2011) reported that "ambitious schedule estimates" were one of several proximal causes of cost growth in the DDG-1000 and JSF programs.
- Bodilly (1993) determined that LANTIRN program plans included ambitious goals that were, in part, responsible for the program's schedule slips. These goals, motivated by the need to fulfill an urgent requirement, included technical ambitions far beyond those that would have been supported by the technical base at the time.
- According to GAO (1986a, pp. 6–7), in the Army's accelerated Sergeant York air defense system acquisition program, "Technical difficulties can be encountered in developing any complex system, irrespective of the acquisition strategy used. However, the Sergeant York's tight schedule and the limited operational testing, both of which were critical elements of the strategy, left few opportunities to resolve these difficulties before major production commitments were made."

Personnel issues have also reportedly caused schedule slips. Individuals with whom we spoke asserted that details about personnel issues are not publicly releasable because of personnel restrictions and thus are not well documented. However, Bodilly (1993) found that schedule slips in the Air Force's LANTIRN program were caused not only by ambitious requirements, as discussed above, but also by extensive concurrency and an inexperienced system program office. Finally, Cashman (1995) identified delays at the subcontractor level, delays in getting necessary data, manufacturing problems, and staffing problems as four of the five most common sources of schedule growth.

Acquisition Approach

The acquisition system utilizes several practices that, according to some reports, may optimize technical performance or cost over cycle time or schedule growth.

Competition

Promoting effective competition is one of the main tenets of the BBP initiative (see Carter, 2010b) and has been stressed in nearly all acquisition reform initiatives over the past 40 years as a means for achieving best value and innovation. The benefits of competition are widely publicized but do not necessarily include shorter schedules. In describing the benefits of competition, Birkler et al. (2001) cited improved product quality, lower unit costs, technological progress, and improved industrial productivity. While improved productivity and technological progress may lead to shorter schedules, Birkler et al. (2001) have suggested that the competitive process requires "additional time and money and entails extra management complexity and effort" (p. 29). Other studies have also suggested that competition may lengthen schedules (Drezner and Smith, 1990, p. 43; Tyson et al., 1989, p. VII-7).

Undefinitized Contract Actions

OUSD(AT&L)'s *Performance of the Defense Acquisition System: 2013 Annual Report* found that UCAs in development (but not in early production) "had a measurable increase on total contract cost growth and also on cycle time in development . . . . A contract with a UCA generally lasted 0.3 years longer (all other things being equal), as measured across all DoD contracts from 1970–2011. Thus, UCAs increase development cycle time by increasing schedule growth" (OUSD[AT&L], 2013, p. 54).

Tailoring the Acquisition Process

DoD Instruction 5000.02 on the operation of the defense acquisition system allows the tailoring of program requirements—such as reporting and documentation—to meet the needs of a given program (DoDI 5000.02, 2008). This includes programs whose products must be acquired more rapidly than normal. However, research suggests that tailoring acquisition may be difficult at times and not well incentivized (Drezner et al., 2011). Anecdotal reporting gathered by Drezner et al. (2011) suggests that process tailoring is sufficient to address the unique requirements of ships, but many others have said that they found tailoring difficult. Instructions dating from 2008 may be insufficient or unclear, and much of the tailoring burden is handled through informal processes (such as coordination) and is therefore difficult to recognize, measure, and anticipate (Drezner et al., 2011). Other research has reported that there can be few incentives for the program manager to tailor a program's acquisition strategy to take on the risk associated with streamlining parts of the acquisition process, noting a lack of tailoring guidance and specific training to implement such guidance. In the 2010 Army Acquisition Review, Decker et al. (2011) found that the Army acquisition system at the time allowed for tailoring but then required each program to "perform all the steps and produce all the documentation of the most complex, technically challenging development," effectively making tailoring infeasible. In the case of information technology programs, a Defense Science Board Task Force found that tailoring in the extant acquisition process has not produced sufficiently shorter timelines (DSB, 2009a). In response, the task force for the acquisition of information technology (IT) recommended a new acquisition process for IT systems to keep up with the rapidly advancing industry (DSB, 2009a). Better Buying Power (BBP) 2.0 recognizes this critique and tries to promote better use of tailoring, and a forthcoming update to the DoDI 5000.02 may include additional support for tailoring.

Other Factors Outside of Program Management Control

Drezner and Smith identified that external guidance (such as OSD or congressional direction, reviews, restrictions, and designations) and events (such as inflation, strikes, and natural disasters) cause delays on the order of one year and that increased program stability (funding, requirements, and guidance) leads to better schedule performance (Drezner and Smith, 1990, p. 43).

CHAPTER THREE

Improving Schedule Performance

In this chapter, we present specific analyses and program experiences that have claimed to provide strategies and mechanisms for improving schedule performance.[1] We present a broad overview of possible solutions offered by the literature and then focus on individual areas of schedule performance that experts have specifically identified for improvement.

Successful Practices That May Not Be Applicable to All Programs

The suggested mechanisms for improving schedule performance follow from the identified causes of schedule slip. The most common recommendations for reducing cycle time are strategies to manage technical risk. One recommendation is to use incremental fielding and evolutionary acquisition (EA) strategies, when appropriate, and to develop derivative products rather than brand-new designs (see, for example, Interim DoDI 5000.02, 2013; Boehm et al., 2010; GAO, 2003). Other possible approaches include maintaining stable funding (GAO, 2010; Kassing et al., 2007; Johnson, 1999; OUSD(AT&L), 2013) and using tailored contracting vehicles, as appropriate (GAO, 2012c; Greenhouse, 2000; OUSD(AT&L), 2013). A summary of the broad range of approaches to improving program performance is provided in the

[1] In this report, we do not examine whether these strategies will work for programs or under what specific conditions; rather, we report what is available in the literature.

GAO report *Defense Acquisitions: Strong Leadership Is Key to Planning and Executing Stable Weapon Programs* (GAO, 2010, p. 1). The GAO lists various characteristics of programs that were on track with original cost and schedule goals. Common themes in these programs—such as having senior leadership support, program priority, program stability, incremental fielding and EA strategies, mature technology, and realistic schedule estimates—are confirmed elsewhere in the literature; however, caution needs to be exercised when applying these strategies because they do not work necessarily for all programs.[2] In fact, there are strategies (for example, prototyping and concurrency) that can either cause schedule growth or shorten schedules, depending on the unique circumstances of each acquisition program and how well the approach is applied. According to GAO, "These practices are in contrast to prevailing pressures to force programs to compete for funds by exaggerating achievable capabilities, underestimating costs, and assuming optimistic delivery dates" (GAO, 2010, p. 1).

GAO has documented other ways to potentially improve schedule performance, dating back to reform initiatives from the 1980s. These include some of the approaches mentioned above, in addition to

> acquiring weapons in larger, more economical quantities, . . . [providing] adequate front-end funding for test hardware, . . . [and] having enough test versions of the weapon to permit concurrent rather than sequential testing of performance, reliability, and other characteristics. (GAO, 1986b, pp. 5–6)

There are specific programs that, in the recent past, were able to rapidly deliver capabilities. Perhaps the most famous and largest such program is the MRAP. This program successfully fulfilled an urgent operational need (UON), both through the effective employment of many of the strategies discussed in this chapter and through a unique set of circumstances, including rapid access to large amounts of budgetary resources and significant senior leadership and congressional support. This unique set of circumstances is rare. Thus, while the MRAP expe-

---

[2] For instance, there may be negative results if senior leadership is pushing to field a program for which the technology is not mature.

rience provided some lessons learned, replicating the MRAP approach on more typical programs may not be feasible. However, it is a good example of how DoD is able to successfully prioritize schedule above cost and many aspects of technical performance.[3] A detailed case study of the MRAP program is presented in the appendix.

Table 3.1 provides a summary of ways to improve schedules that have been suggested in the literature.

Requirements Development, Generation, and Management

To address negative schedule effects resulting from the requirements process, various reports promote the benefits of stable, realistic requirements while recognizing that military needs change over time. As discussed in Chapter Two, instability in requirements is strongly related to cost and schedule growth and, thus, should be avoided as much as possible. However, maintaining responsiveness to military threats may require flexibility in requirements.

The Navy's Poseidon P-8A program—viewed by many as a success—illustrates the importance of feasible and stable requirements. The P-8A stayed on schedule, at least partially because of stable and technically feasible requirements (e.g., through an emphasis on using proven technologies), the use of a commercial derivative, early advanced development efforts, and effective program management (Blickstein et al., 2011, p. 8).

Farr, Johnson, and Birmingham (2005) have suggested that working closely with the user community helps to manage expectations as a means to maintain program schedules. They also acknowledged that requirements must, at times, evolve and be flexible.[4] According to Brodfuehrer (2000), requirements can be better managed by working

---

[3] The MRAP is unique case, but the cost of the program has been high and operations and support performance very low, so many vehicles are now being scrapped as a result.

[4] The use of configuration steering boards in BBP 2.0, promoted primarily for their affordability, can also be used to control or even reduce requirements to maintain a schedule.

Table 3.1
Possible Ways to Improve Schedules in the Acquisition Literature

Area	Possible Ways to Improve Schedules	Analysis	Current Legislation and Policy (selected)
Requirements development, generation, and management	Stable and realistic requirements	GAO, 2010, p. 1; Blickstein et al., 2013; OUSD(AT&L), 2013, pp. 33–35	*
	Collaboration between the program management and end-user communities (with proper management)	Brodfuehrer, 2000, pp. 22–27; Farr, Johnson, and Birmingham, 2005	Kendall, 2013
	Proper management of flexible requirements	Farr, Johnson, and Birmingham, 2005	Kendall, 2013
Managing technological risk in development and production	Use of mature/demonstrated technology to ensure a high level of maturity before production	GAO, 2010, p. 1; DSB, 2009b; GAO, 1999, pp. 3–4; GAO, 2010, p. 1; Drezner et al., 2011; DoD, 2001, p. 49; OUSD(AT&L), 2013, pp. 33–35	Kendall, 2013
	Use of incremental fielding or EA strategies and the development of derivative products (rather than brand-new designs)	GAO, 2010, p. 1; Johnson, 1999; Brodfuehrer, 2000	Interim DoDI 5000.02, 2013
	Employment of "agile" methods that easily respond to changes in software development	Lapham et al., 2010	*
	Prototyping	Tyson et al., 1989; Drezner and Huang, 2009	DoDI 5000.02, 2008; Weapon Systems Acquisition Reform Act of 2009 (WSARA) (Pub. L. 111-23, 2009)

Table 3.1—continued

Area	Possible Ways to Improve Schedules	Analysis	Current Legislation and Policy (selected)
Managing technological risk in development and production (continued)	Concurrency in programs with low technological risk	Howitz, 2008, pp. 32–33; Kendall, 2012b; GAO, 2012a, p. 52; Younossi et al., 2005, p. 56	Kendall, 2012b
	Use of commercially derived items	DoD, 2001, p. 49	DoDI 5000.02, 2008
	Use of the commercial practice of fixing the design before the production contract award	Comptroller General of the United States, 1979, p. 6; GAO, 2009, p. 1; Kendall, 2012b	DoDI 5000.02, 2008; WSARA (Pub. L. 111-23, 2009)
	Use of the commercial practice of reducing the design's complexity	Comptroller General of the United States, 1979, p. 6	DoDI 5000.02, 2008
Resource allocation	Stable funding	GAO, 2010, p. 1; Kassing et al., 2007; Johnson, 1999; OUSD(AT&L), 2013, pp. 33–35	*
	Adequate test funds (hardware, modeling and simulation)	GAO, 1986b, p. 5; Farr, Johnson, and Birmingham, 2005	*
Acquisition management: internal to the program	Bypassing competition (including employing multiyear or sole-source procurement strategies in the production phase)	Tyson et al., 1989, p. VII-7; Gailey, 2002; Reig, 1995	DoDI 5000.02, 2008; Carter, 2010b; Kendall, 2013
	Preplanned product improvement	GAO, 1986b, p. 5	DoDI 5000.02, 2008

Table 3.1—continued

Area	Possible Ways to Improve Schedules	Analysis	Current Legislation and Policy (selected)
Acquisition management: internal to the program (continued)	Acquisition of the same number of units but in larger, more economical quantities in the production phase	GAO, 1986b, p. 5	Carter, 2010a, 2010b
	Emphasis and adherence to schedule as a program priority (e.g., MRAP program)	Ward, 2006; Brodfuehrer, 2000	*
	Development and maintenance of a comprehensive and realistic master schedule	GAO, 2013, pp. 7, 21; Anderson and Upton, 2012, p. 36	DoDI 5000.02, 2008
	Use of contracting vehicles to expedite contracting process (existing contracts, undefinitized contracts in low-rate initial production [LRIP], sole-source contracts)	GAO, 2012c; Greenhouse, 2000; OUSD(AT&L), 2013, pp. 33–35	Carter, 2010a, 2010b; Kendall, 2013
	Operational testing and evaluation (OT&E) results available before production startup	GAO, 1990, p. 1	DoDI 5000.02, 2008
	Use of modeling and simulation to reduce the risk and duration of live tests	Fox et al., 2004	DoDI 5000.02, 2008
	Involvement of the test community in all program phases	Farr, Johnson, and Birmingham, 2005	DoDI 5000.02, 2008
	Use of integrated product teams	DoD, 2001, p. 49	DoDI 5000.02, 2008
	Improved program stability in general, including funding and requirements	GAO, 1986b, p. 5; OUSD(AT&L), 2013, pp. 33–35	*

Table 3.1—continued

Area	Possible Ways to Improve Schedules	Analysis	Current Legislation and Policy (selected)
Acquisition management: internal to the program (continued)	Realistic schedule estimates	GAO, 2010, p. 1; OUSD(AT&L), 2013, pp. 33–35	WSARA (Pub. L. 111-23, 2009)
Acquisition management: external to the program	Senior leadership support	GAO, 2010, p. 1	*
	Program identified as a priority	GAO, 2010, p. 1	*

* Lack of an entry does not imply that these techniques are not being employed. The completed cells in this column provide references to selected implementations that were identified in our search.

closely with the end user early in the development process. However, he also pointed out that flexible requirements can be problematic. If the government is not careful, lack of specification in requirements could lead to the development of an undesirable product. In addition, if end-user involvement is not managed properly, the requirements development process may yield an unwieldy set of requirements.

Resource Allocation

As summarized in Chapter Two, unstable or inadequate funding may result in longer schedules (Kassing et al., 2007; Drezner and Smith, 1990). Understandably, then, experts have suggested maintaining the stability and adequacy of funds as part of the recipe for improving program schedule performance (Johnson, 1999). A study investigating successful approaches to rapid acquisition identified early access to adequate funding as necessary to support the development of an acquisition strategy, test and evaluation strategy, and other programmatic documentation that speeds execution (GAO, 2012c). John C. Wilson, former director of system acquisition in what is now OUSD(AT&L), asserted that program managers should "advocate and seek as fully and completely as possible the funding that will allow a program to be quickly and efficiently executed" (quoted in Johnson, 1999, p. 11).

The literature offers few proven strategies for improving program funding stability and adequacy. One possibility is advanced appropriations (Blickstein and Smith, 2002),[5] but its limited flexibility for management and Congress to respond to unplanned events makes it difficult to achieve. Thus, it is unclear whether advanced appropriations will become widespread.

Former MRAP program manager Thomas Miller (2010) believes that lessons learned from the MRAP experience can be applied to more typical acquisition programs to lessen the effects of funding instability on individual programs. First, decisionmakers should keep program managers informed of potential funding cuts. Second, decisionmakers should consider cutting low-value programs (a portfolio view) rather than spreading cuts across all programs (the latter tactic is often known as a "peanut butter spread" or "salami-slice" approach). Of course, these changes will not make funding stable for low-priority programs, and advanced notice of potential cuts may not make much difference if few hedging options are available.

Managing Technical Risk in Development and Production

The studies that have focused exclusively on schedule growth in weapon system programs have identified technical difficulties as having the most profound impact on program length. For example, in 2005, the *Defense Acquisition Review Journal* published a literature review of nearly a dozen studies describing sources of schedule slippage (see Monaco and White, 2005). The most commonly cited source of schedule slippage was technical in nature. Drezner and Smith (1990) and Cashman (1995) also identified technical difficulties as having the most profound impact on program length.

---

[5] Under advanced appropriations, funds are not appropriated for a single year but instead are distributed over multiple years and available in each of the following years.

The literature identifies many strategies for mitigating technical risk. We present selected highlights in the sections that follow.

Starting with Mature Technology and Design

Many reports have suggested that starting with mature technology will help minimize the likelihood of schedule problems (GAO, 1999, p. 5; GAO, 2010, p. 1). There are many examples of streamlined or expedited acquisition processes that were enabled by the use of mature technologies or commercially available, nondevelopmental items. A U.S. Army Audit Agency report (2011) described how this approach resulted in many programs entering the acquisition process at Milestone C,[6] streamlining or bypassing many of the earlier acquisition requirements, but noted delays resulting from a lack of additional tailoring and staffing of capability documents. In an earlier report reviewing TRLs as applied to 23 technologies,[7] GAO concluded that those programs with a higher level of technical maturity before incorporating new technology were in a better position to succeed (GAO, 1999, pp. 3–4). According to the same report, "Programs with key technologies at readiness levels 6 to 8 at the time of program launch met or were meeting cost, schedule, and performance requirements" (GAO, 1999, p. 4). In recent testimony, Kendall (2012b) mentioned that TRLs are useful but often not sufficient for understanding technology risk. Chittenden has suggested that ensuring that program technology "is mature enough to realize fielding within five years of the program's Milestone A decision" is one way to alleviate the concerns associated with weapon system complexity and schedule duration (Chittenden, 2012, p. 103).

Appropriate knowledge of the state of the technology, design, and manufacturing process is also said to influence program success. In a 2009 study, GAO identified programs that it claimed proceeded with insufficient knowledge, noting, "For 47 programs GAO assessed in-depth, the amount of knowledge that programs attained by key deci-

---

[6] Acquisition programs typically enter the acquisition process at Milestone A (design) or Milestone B (development) rather than Milestone C (production).

[7] A TRL is a DoD measure used to assess the maturity of evolving technologies during development.

sion points has increased in recent years; but most programs still proceed with far less technology, design, and manufacturing knowledge than best practices suggest and face a higher risk of cost increases and schedule delays" (GAO, 2009, p. 1).

In addition to ensuring sufficient maturity in technology, proceeding with a mature design may improve schedule performance. In an older report, the Comptroller General of the United States cited a practice in the commercial sector that prevented schedule growth: "The commercial practice of fixing the design before contract award, along with less complexity involved with the design, has contributed greatly to the commercial sector's success at holding down cost and schedule overruns" (Comptroller General of the United States, 1979, p. 6). Of course, there are multiple contract award points in defense acquisition, but current practices recognize the importance of completing the preliminary design review before Milestone B (DoDI 5000.02, 2008; Pub. L. 111-23, 2009). The prior discussion of concurrency addressed the trade-offs in starting production before designs are complete (see, for example, Kendall, 2012b).

Incrementally Fielded or Evolutionary Acquisition

Incremental fielding and EA are acquisition strategies that have been employed as a way to speed fielding and control technical risks. They aim to provide some initial operationally useful capabilities more quickly than processes that use a single step to acquire a capability. EA achieves this goal through incremental improvements, which are less demanding than those typically seen through the traditional process.

For example, current DoD acquisition policy designates incremental fielding of software-intensive programs as one possible approach to consider:

> Incrementally Fielded Software Intensive Program . . . is a model that has been adopted for many DBS [Defense Business Systems]. It also applies to upgrades to some command and control systems or weapons systems software where fielding will occur in multiple increments as new capability is developed and delivered, nominally in 1- to 2-year cycles.

This model is distinguished from the previous model by the rapid delivery of capability through several limited fieldings in lieu of single Milestones B and C and a single full deployment. Each limited fielding results from a specific build, and provides the user with mature and tested sub-elements of the overall capability. Several builds and fieldings will typically be necessary to satisfy approved requirements for an increment of capability. The identification and development of technical solutions necessary for follow-on capabilities have some degree of concurrency, allowing subsequent increments to be initiated and executed more rapidly.

This model will apply in cases where commercial off-the-shelf software, such as commercial business systems with multiple modular capabilities, are acquired and adapted for DoD applications. An important caution in using this model is that it can be structured so that the program is overwhelmed with frequent milestone or fielding decision points and associated approval reviews. To avoid this, multiple activities or build phases may be approved at any given milestone or decision point, subject to adequate planning, well-defined exit criteria, and demonstrated progress. An early decision to select the content for each follow-on increment (2 through N) will permit initiation of activity associated with those increments. Several increments will typically be necessary to achieve the required capability. (Interim DoDI 5000.02, 2013, p. 11)

According to some published EA best practices:

Think "parallel developments:" Often in an evolutionary model, development of increments must occur in parallel to deliver capability on time. Increments may vary in time to develop and integrate. If done serially, they can extend the program schedule and adversely impact the ability to deliver capability to the users in a timely manner, which was the purpose of evolutionary acquisition. Managing parallel development is challenging but not unachievable; it should not be avoided. Make use of configuration management to control the development baselines and track changes. Allow time in the increment development schedules for the reintegration of a "gold" baseline for final incorporation of parallel changes prior to test and fielding. (MITRE, 2012)

Prototyping and Testing

Prototyping can help mature the technology being employed, and testing can measure the performance and determine the level of remaining risks. These approaches reportedly carry schedule implications as well.

The relationship between prototyping and schedule reported in the literature is varied. In an older evaluation of cost and schedule growth of programs that have involved prototypes, Tyson et al. (1991) concluded that prototyped programs help predict development costs, but these programs were somewhat longer in duration than those that did not involve prototyping. The authors did not identify the reasons why these programs took longer, but they hypothesized that the increased length may have been due to the technical complexity of the programs studied.

OT&E activities have also been identified as an area that may yield schedule improvements. A GAO report from the 1980s described two initiatives of the then–Deputy Secretary of Defense to reduce acquisition timelines and mitigate schedule slippage, including "emphasizing preplanned product improvement and obtaining adequate funding for test hardware" (GAO, 1986b, p. 5). This latter recommendation was in response to a previous GAO report concluding "that test schedules of weapon systems were constrained, in part, by too few prototypes available for testing. As a result, expensive retrofits were required to correct problems identified during operational testing performed after the production decision was made" (GAO, 1986b, p. 6).

GAO's review of 33 acquisition improvement initiatives reported on the importance of meeting performance requirements "in a representative operational environment" prior to beginning production. The GAO determined that many of the major systems reviewed did not follow this best practice and, as a result, they required expensive retrofits following operational testing. The experience of the Sergeant York program is an extreme example. The program "was canceled after 64 systems costing $1.8 billion had been produced and delivered because, according to the Secretary of Defense, independent operational tests showed that the system's performance did not meet the military threat" (GAO, 1986d, p. 29). In another report, GAO also claimed that "making OT&E results available before production start-up could help preclude cost growth, schedule slippages, and performance shortfalls that frequently arise during the later phases of a weapon system's development" (GAO, 1990, p. 1).

The literature identifies potential improvements to testing processes and procedures. Bernard Fox and colleagues (2004) suggested that the use of integrated contractor-government test teams and a reevaluation of test procedures could optimize testing. This same report (p. xxi), along with Farr, Johnson, and Birmingham (2005), described modeling and simulation as techniques that can reduce risk and duration of tests. Farr, Johnson, and Birmingham also found that the test community may delay fielding until a 100-percent solution is achieved, concluding that involving the test community in all phases of a program would help maintain schedules by ensuring that issues are uncovered as soon as possible. However, Fox and colleagues point out that staffing limitations may preclude early integration of testing and evaluation personnel (B. Fox et al., 2004, p. xxi).

Development and Manufacturing Approaches

As discussed earlier, concurrency has been identified as a possible means for reducing program cycle time. If technical risk is low and concurrency is managed and executed properly, concurrency can result in reduced schedules. If technical risk is too great or not managed and executed properly, then concurrency can lead to significant cost and schedule growth. The MRAP program, for example, employed proven, mature, and commercially available technologies and was therefore successful in implementing concurrency (see the appendix). Moreover, the question is how much concurrency is the right mix of technical, cost, schedule, and operational capability risk (see, for example, Kendall, 2012b).

The literature cites a host of development and manufacturing approaches that may improve schedule performance. Many of these approaches are found in the software development world. For example, agile architectures[8] and service-oriented architectures[9] are cited as approaches that can help improve schedule performance for software development activities. In a 2009 study, the Software Engineering Institute at Carnegie Mellon University published a report suggesting that, while some challenges must be overcome, DoD could employ agile methods to procure software.[10] Proponents of agile architectures argue that this development approach delivers software capability more quickly. In a 2009 briefing on reducing the acquisition cycle time in technology insertion, Bliss discusses broader agile approaches, stating, "Structuring the DoD enterprise for agility in responding to rapidly developing and constantly changing environment is . . . At least as important as investing [in] new technology" (Bliss, 2009, slide 15). In addition to concurrency and agile acquisition, inserting "commercial parts and technology in weapon systems" (Lorell and Graser, 2001, p. xxiii) and implementing lean manufacturing or competitive manufacturing have also been identified as potentially effective ways to reduce schedules (see, for example, Neves and Strauss, 2008, p. 22).

---

[8] According to Yakyma and Leffingwell (2013), the "Seven Principles of Agile Architecture" are as follows:

1. Design emerges. Architecture is a collaboration.
2. The bigger the system, the longer the runway.
3. Build the simplest architecture that can possibly work.
4. When in doubt, code, or model it out.
5. They build it, they test it.
6. There is no monopoly on innovation.
7. Implement architectural flow.

[9] According to IBM, service-oriented architecture has the following characteristics: It enhances the relationship between enterprise architecture and the business, it allows developers to build composite applications as a set of integrated services, and it provides flexible business processes (Portier, 2007).

[10] Agile is a philosophical development approach. For more information, see Lapham et al., 2010.

Defense Acquisition Management: Practices, Policies, and Procedures

DoD's *Guide to the Project Management Body of Knowledge* has recommended the use of "realistic cost estimates and fully funding those estimates" as a best practice (DoD, 2003). This is also a requirement for Milestone B certification: "[R]easonable cost and schedule estimates have been developed to execute, with the concurrence of the Director of Cost Assessment and Program Evaluation, the product development and production plan under the program" (10 U.S.C. § 2366b). Incremental fielding is one possible model for software-intensive programs in current DoD policy (Interim DoDI 5000.02, 2013). Additionally, commercial items can, in some cases, have further benefits. These and other acquisition strategies are summarized below.

Program Management: Schedule Estimation and Management

Neves and Strauss (2008, p. 21) have asserted that adhering to an aggressive schedule requires making trade-offs:

> All programs have a measure of schedule pressure. Once baselined, the 'iron triangle' of cost, schedule, and technical scope is at play. But truly schedule-driven development programs behave differently and have different needs. Attempting to plan, execute, and manage a truly schedule-driven development effort as if it were a standard acquisition program done faster will not work, will slip, will cost more—and will probably get you fired.

For example, GAO questioned how MRAPs would fit into the services' long-term plans, suggesting that the services dispensed with the typical discussions regarding the long-term utility of the vehicle in the interest of delivering the capability as quickly as possible (Sullivan, 2009). Reliability, mobility, and safety challenges surfaced following deployment; these may not have been issues if the program had been given more time to plan, develop, and produce a better-engineered vehicle (Sullivan, 2009). Additionally, and perhaps most importantly, because of immediate wartime needs, decisionmakers prioritized the MRAP's schedule above potential technical improvements, and the user community was "willing to accept a useful solution in the short term while the program management office continued to develop the system to its desired end-state" to deploy a basic capability as quickly as possible (Blakeman, Gibbs, and Jeyasingam, 2008, p. 19). In another example, the F/A-18E/F program was able to control cost and schedule growth by limiting performance requirements more effectively than other programs. While the F/A-18E/F provided "some incremental improvements over the C/D model, especially in the areas of stealth, range, and payload capacity . . . the [program] sought lower performance in some areas compared with the F-14" (Younossi et al., 2005, pp. 4–5). A major issue for program management is whether "revolutionary" capabilities and technologies are needed to meet the threat. Revolutionary new technologies and capabilities will involve inherently risky development programs. A key question is how development programs seeking revolutionary breakthroughs can best be managed.

Other analysts have suggested that schedules may be improved through increased management prioritization and schedule oversight. Ward and Quaid (2006) recommended setting aggressive schedule improvement goals for all DoD programs and tracking their schedule performance with cycle-time metrics. They also suggested introducing schedule incentives into more DoD contracts. Brodfuehrer (2000) emphasized tracking progress and sustaining urgency.

The literature also highlights the importance of personnel quality and effective communication in managing schedules. According to Brodfuehrer (2000), ensuring that the program has the right staff and that there is a good understanding of how the system will evolve over time could minimize disruptions caused by upgrades, changing technology, or new requirements. Farr, Johnson, and Birmingham (2005) have emphasized the importance of good and continuous communication among members of the development team to maintain the schedule, including the program manager, contractor, and government staff. Anderson and Upton (2012) have promoted the use of an IMS to manage and control schedule, rather than simply treating it as an unused oversight document. Specifically, "there is a colossal gap in available resources skilled in the management and use of the IMS. All these factors drive the underutilization of the IMS, which plays a large part in the plague of program cost overruns and late deliveries" (Anderson and Upton, 2012, p. 36).

Acquisition Approach

Contractor Incentives and the Contracting Process

Providing the right incentives could help to improve contractor schedule performance (Schinasi, 2008, p. 2). In an evaluation of shipbuilding in the United Kingdom, Arena et al. (2005) reported that commercial contracts employed more incentives for on-time delivery than their military counterparts during their review period, and these incentives contributed to better on-time performance. DAU currently provides guidance on using incentives with schedule goals:

- When assigning a profit/fee value for technical risk, consider "above normal value" when "[t]he contractor has accepted and accelerated delivery schedule to meet DoD requirements" (OUSD[AT&L], 2012).
- In contracts, incentives can be used to motivate schedule or delivery, along with performance and cost incentives (OUSD[AT&L], 2012).

As discussed in Chapter Two, competition may induce additional incentives, but its ability to reduce cycle times is unclear. In fact, the majority of the literature on the relationship between contract strategies and schedule supports sole-source and multiyear procurement as the most schedule-friendly approaches.[11] For example, Tyson and colleagues reported that competition led to longer programs, while multiyear and sole-source procurements led to reduced program lengths (Tyson et al., 1989, p. VII-7). Moore and colleagues (2012) identified sole-source contracts as providing a quicker way to satisfy an urgent customer requirement. In an assessment of the F-22 program, Yonoussi et al. (2007, p. xv) found benefits in multiyear procurement, such as the ability to "schedule workers and facilities more efficiently

---

[11] Sole-source contracting is subject to various laws and regulations, which must be taken into account.

and reduce the burden of preparing multiple proposals," which could, in turn, yield schedule improvements.

Neves and Strauss (2008) warned that objective award fees based on delivery dates can prioritize schedule performance at the expense of other factors, increasing the probability of on-time delivery but also risking compromised technical and cost discipline. They advised using multiple balanced incentives to avoid incentivizing such undesirable side effects. Neves and Strauss also suggested developing a risk plan to explicitly manage competing objectives, offering the following advice to program managers:

> Push your contractor—and yourself—to actually develop the risks and their mitigation plans. . . . Risk plans that merely exist in presentation material and have not been developed so that schedule, performance, and cost impacts are known in terms of the program integrated master schedule, system specification, test plans, and development capacity are worse than having no risk management at all. (Neves and Strauss, 2008, p. 23)

Many sources have identified undefinitized contracts as an additional technique to make progress while the contracting process is finalized. However, according to OUSD(AT&L) (2013, p. 54), "Undefinitized contract actions (UCAs) had a measurable increase on total contract cost growth and also on cycle time in development" on MDAP contracts from 1970 to 2011. Interestingly, the office also reported that "UCAs did not correlate with total cost growth in early production as they did on development contracts" (p. 62). This seems to indicate that UCAs may be successfully employed in early production to help maintain schedules, and that UCAs may not be appropriate in all parts of the acquisition process.

Another contracting option, indefinite-delivery/indefinite-quantity (IDIQ) contracts, might also streamline schedules. This recommendation mainly applies to the production phase and lower-tier parts and component suppliers. According to the General Services Administration, IDIQ arrangements "help streamline the contract process and speed service delivery" (Thompson, 2013). Greenhouse describes other benefits of IDIQ contracts, including flexibility in stated limits, increased contracting efficiency, lower contracting costs, and lower proposal costs (Greenhouse, 2000, p. 18).

Tailoring the Acquisition Process

As discussed in Chapter Two, the DoD 5000-series guidance documents allow programs to tailor the acquisition process and requirements documentation; however, for multiple reasons, program managers are often reluctant to take advantage of tailoring options.[12] In the latest guidance recently released, tailoring is given greater focus:

> The structure of a DoD acquisition program and the procedures used should be tailored as much as possible to the characteristics of the product being acquired, and to the totality of circumstances associated with the program including operational urgency and risk factors. (Interim DoDI 5000.02, 2013)

Weapon system programs have received waivers when it comes to many of the requirements dictated by the Defense Acquisition System. As a pilot program, the Joint Direct Attack Munition program implemented provisions of the Federal Acquisition Streamlining Act, which allowed the program "to use commercial item exemptions for noncommercial items" (Myers, 2002, p. 316). These waivers and exemptions streamlined the program's review processes and reporting procedures, contributing to the on-time delivery of munitions (Myers, 2002).

Rapid Acquisition and Urgent Operational Needs

The majority of the reviewed literature describing schedule growth and improvement refers to schedule growth observed in "traditional" acquisition programs. By contrast, DoD also procures materials for UONs or joint urgent operational needs (JUONs). These programs are a high priority and involve capabilities that must be delivered to the warfighter quickly. Lessons learned from UON and JUON acquisition programs can help illuminate the challenges that DoD faces with rapid acquisition or accelerated schedules. A Defense Science Board

---

[12] The *Defense Acquisition Guidebook*, in particular, emphasizes the tailoring of processes and documentation (see DAU, 2013).

task force reported in July 2009 that DoD "lacks the ability to rapidly field new capabilities for the warfighter in a systematic and effective way" (DSB, 2009b, p. vii). The task force recommended establishing a process to determine which programs needed a rapid acquisition process and developing a unique process to efficiently and effectively satisfy these urgent needs. The task force suggested establishing a new agency to implement this rapid process and that the executive and legislative branches should establish a separate funding stream to satisfy urgent needs. Finally, it emphasized that "any rapid response must be based on proven technology and robust manufacturing processes," underscoring the importance of proven, mature technology and processes, regardless of the unique privileges afforded to a program (DSB, 2009b, p. ix). As a result of this concern, the Joint Rapid Action Cell was established within OUSD(AT&L) and is responsible to the Secretary of Defense through the USD(AT&L) and USD(Comptroller). It monitors, coordinates, and facilitates meeting combatant commanders' urgent warfighting needs (DAU, 2008).

A recent GAO report (2012c) stated that some of the challenges identified with rapid acquisition programs involve the lengthy process of getting a contract awarded. Among the initiatives that GAO studied, those that were successful used existing contracts, undefinitized contracts, sole-source contracts, and other tools to help expedite contracting. Off-the-shelf initiatives took longer than other initiatives to get on contract because they did not have the option of leveraging existing contracts. However, they were able to field capabilities more quickly once their contracts were awarded (GAO, 2012c).

The MRAP program is one of the most recent successful major programs to implement rapid acquisition. The program was able to deliver vehicles at a significant rate less than two years after its inception (GAO, 2009). Several conditions enabled this quick production and fielding, including high-level leadership support and a special high-priority status within DoD, which facilitated some of the program's nontraditional acquisition approaches. The program also utilized creative contracting strategies—including undefinitized contracts—and was committed to the use of proven technologies. A detailed case study of the MRAP program is provided in the appendix.

CHAPTER FOUR
Conclusions

Given the continuing interest in ensuring that acquisition program cycle times and schedule growth are reasonable and minimized, we reviewed and summarized the recent and historical literature on the two issues involving program schedule: schedule slip and longer program schedules. The open-source literature includes a range of program examples, quantitative and qualitative analysis, expert opinions, and conceptual assertions. Our review did not attempt to judge the strength of the evidence supporting these assertions and analyses but, rather, surveyed as broadly as possible the full range of causes put forward by experts in the field. We also note that reports may reflect different conditions over time or differences between sampled programs. We identified the following reasons for schedule delays in the literature that we reviewed:

- The reason asserted most often was **difficulty managing technical risk** (e.g., program complexity, immature technology, or unanticipated technical issues).
- The second most common reason was **initial assumptions or expectations that were difficult to fulfill** (e.g., schedule estimates, risk, requirements, or performance assumptions).
- Another common reason was **funding instability**, which complicates management and can directly stretch production schedules.

Some of these processes and activities occur outside of the control of program management, while others fall under its control. The literature describing sources of schedule growth tends to focus on negative, rather than positive, program examples. Thus, many recommendations for schedule improvement in the literature may be based more on avoiding pitfalls than on following proven paths to success.

The most common recommendations to reduce cycle time and control schedule growth are strategies for better managing technical risk. These recommendations include using incremental fielding and EA strategies and developing derivative products (rather than brand-new designs), as well as using mature or proven technology (i.e., commercially available components). Other recommendations for mitigating issues beyond technical risk include maintaining stable funding, using atypical contracting vehicles, and making the schedule a top priority (e.g., at higher expense or lower technical requirements). Some strategies identified require careful consideration of program characteristics prior to program initiation; otherwise, schedules may suffer as a result. No strategy is appropriate in every case, so careful judgment and balancing are required. For instance, prioritizing the schedule may be challenging in a program with immature technology that has no options for incorporating mature technology.

The schedule is intrinsically tied to a large set of broader considerations for program managers and other stakeholders. The complexity of these relationships makes it difficult to isolate the causes of prolonged cycle times and schedule growth. It also makes testing mitigation strategies for reducing cycle times and stemming schedule growth problematic. However, if common themes in the literature are valid, then **progress in the following areas may yield schedule improvements**:

- reducing technological risk
- providing realistic estimates and expectations
- ensuring the feasibility and stability of requirements.

We address each of these considerations in turn, along with related drawbacks and potential directions for future research.

Reducing Technological Risk

If the schedule is to be prioritized or fixed in the "iron triangle" of cost, schedule, and technical performance, the literature asserts that managing or reducing technical risk must be a priority. Not one source we reviewed claimed that costs could be reasonably adjusted to accommodate an aggressive schedule in a program with a given set of performance expectations. Instead, restrictions on performance requirements and technical risk have been repeatedly emphasized as the keys to reducing timelines and avoiding schedule slippage.

One of the most important aspects of technical risk is the maturity of the technology being used and the complexity of system integration. Most of the programs highlighted in this report for their excellent schedule performance (e.g., P-8A, F/A-18E/F, and MRAP) can be reasonably characterized as evolutionary, lower-capability systems and employed technologies that were relatively proven and in use either in other military weapon systems or in the commercial sector. Those programs highlighted for their poor schedule performance (e.g., F-22 and the Future Combat Systems) are revolutionary systems and were much more ambitious in this respect, requiring (among other things) the development and integration of less mature technologies or new combinations of technologies. While "large leaps" in technological advancement are, at times, desirable and even obtainable, the literature clearly warns against pursuing these leaps under a tight deadline. In cases where revolutionary systems are warranted, managing technical risk would be a priority for program management.

Because it uses incremental (rather than single-step) technological improvements, incremental fielding is cited as one possible model to consider for software-intensive programs (Interim DoDI 5000.02, 2013). Conceptually, it should reduce technical risk substantially by incorporating technologies only after they are sufficiently mature or can be accommodated.

Such concerns may have led to the de-emphasis on EA for hardware-intensive systems since the prior DoDI 5000.02 (2008). Of course, future upgrades to systems can be employed to insert new technologies into systems.

It should be noted here that while incremental fielding should theoretically reduce technical risk and improve schedule performance in specific programs, these improvements do not necessarily translate to accelerated deployment of all desired technologies to the field. Technology must be developed before it can be deployed, and incremental fielding simply recognizes that it is best to insert technology when it is ready rather than forcing it to be ready by fiat. Determining the maturity and operational utility of new technologies prior to Milestone B (when a full-fledged program is established) could save the department time and money. These savings could be achieved not necessarily by reducing time and money spent on fielded systems, but by avoiding significant investments in established programs based on technologies that are not yet ready for incorporation in a system. This cost avoidance could, in turn, free up funds for programs that employ mature technologies and could be accelerated if they had more resources. Unfortunately, analysts have pointed out that placing restrictions on performance expectations and technology insertion in acquisition programs—as encouraged under an EA approach—can run counter to the military's desire to achieve and sustain a competitive edge over its adversaries (Arena et al., 2006). Current processes such as affordability constraints and Configuration Steering Boards (Interim DoDI, 2013; Kendall, 2013) indicate a willingness to make such trades.

The answers to the following research questions would help quantify the problem of technical risk and their costs, allowing these costs to be weighed against the military's desire for faster technology advancement and deployment:

- How much time and money has been spent on programs that were unachievable from the start?
- How many programs in the recent past have been canceled because of technology immaturity?
- How much had been invested in these programs at the time of cancellation?

Improving the Accuracy of Schedule Estimates and Expectations

The literature implies that there is a need to improve schedule estimates and expectations. "Overly optimistic" and "ambitious" are two phrases commonly used to describe the estimates and expectations included in problematic program plans. Overly optimistic schedule estimates blind decisionmakers to the need to make early, informed trade-offs, and they set up programs for later criticism over schedule growth. Correcting inaccuracies and overcoming the pressures associated with program advocacy (a commonly cited source of these inaccuracies) is difficult. Thus, improved schedule estimates could help improve schedules. According to analysts we consulted, schedule-estimating methodologies are not as well developed as cost-estimating methodologies. The limited prevalence of schedule estimation in the literature supports this observation.

The answers to the following related research questions would help quantify the scope of the schedule-estimation problem and offer guidance for improving the tradecraft in DoD:

- To what extent are schedule-estimating skills present in today's acquisition workforce?
  - Where are these skills located? Could they be better employed?
  - How might these skills be fostered?
    - Is better training needed?
    - Should the department seek individuals with more experience in schedule estimating?
- Are program managers incentivized to unreasonably adhere to established schedules? If not, how might incentives encourage better schedule performance?
- How robust are the schedule-estimating methodologies currently used by the acquisition workforce? How might they be improved?
- How does the commercial sector develop and implement schedules for complex high-technology development and production programs? How good is the record in the commercial sector for such estimates?

Improving the Feasibility and Stability of Requirements

The literature asserts that aggressive and rigid adherence to overly ambitious requirements can contribute to prolonged acquisition schedules and schedule growth. Additionally, the literature describes a complicated relationship between requirements instability and schedule growth. According to an OUSD(AT&L) report,

> The time required to acquire next-generation capabilities is often longer than the strategic threat and technology cycles these capabilities are meant to address. Performance (good or bad) in planned defense acquisition is intertwined with cost and schedule implications from unplanned responses to these external demands. This is not an excuse for cost and schedule growth, but an observation from first principles that changing threats and needs can add costs and delays relative to original baselines as ongoing acquisitions are adjusted. (OUSD[AT&L], 2013, p. 109)

However, effective communication among the requirements, acquisition, and test communities has been highlighted as a way to avoid unstable or infeasible requirements (Brodfuehrer, 2000; Farr, Johnson, and Birmingham, 2005; Kendall, 2013). Adherence to a schedule will require that this communication be focused on (1) reducing (rather than expanding) program scope whenever possible and (2) allowing flexibility (rather than enforcing stringency) in performance requirements. This flexibility, at times, means that the user community needs to accept a less-than-perfect solution in the short term, with the understanding that improvements and modifications might be made later on if they are viable and affordable.

Research centered on the following questions may help DoD understand where improvements in this area are needed:

- What has been the experience of technically complex programs with respect to early and continuing user involvement in controlling and adjusting requirements?
  - What is the nature of most interactions between the acquisition and user communities?
    - Are these interactions generally focused on reducing or adding requirements?
    - Have options that involve trade-offs among cost, schedule, and performance been effectively estimated and communicated to users?
  - What types of interactions take place when a requirement is found to be more difficult to meet than originally expected?

        - If a requirement is found to be infeasible, how often is it changed?
        - How long does it take to communicate and make the change?
        - What are the roadblocks to making changes quickly?
        - How have requests for significant changes by the acquisition community been received by users?
    - Are there lessons to be learned from programs in which there was a healthy relationship between the user and acquisition communities?
        - Were there specific program management strategies or techniques that encouraged this relationship?
        - How can positive interactions between these two communities be incentivized?

Closing

While the literature on cost growth over the past 60 years is comparatively rich in data, analysis, and recommendations, the literature on schedule growth is relatively limited. Few studies have looked solely at cycle time and schedule growth, comparing similar data on multiple programs. Even fewer reflect current policies and conditions. Schedule growth tends to be discussed in relation to cost growth for various programs over time. There are multiple research questions that can be pursued, but a broader study of a wide range of major defense acquisition programs, similar to Drezner and Smith's 1990 review of weapon system acquisition schedules, would provide a better understanding of how cycle times and schedule growth have changed over time. The acquisition system would also benefit from comparisons across a large set of MDAPs for which data on schedules for key points after Milestone B (or its equivalent) have been available in Selected Acquisition Reports for decades. In addition, a study of how DoD should best manage very high technology risk programs that ended up being game changers would be beneficial, given that technical risk is a main factor cited in the literature for schedule growth.

The literature documenting potential ways to improve program schedule performance is also relatively thin. Suggestions are typically mixed, with analyses examining other acquisition problems, making it hard to isolate schedule-related factors from other factors. An in-depth analysis focusing specifically on how to improve schedules in acquisition programs is needed. In addition to cross-program analyses, the field could also benefit from additional detailed case studies of individual programs that have successfully stayed on schedule, in addition to the ones that have not. This type of study would help analysts identify how concepts, such as framing assumptions,[1] may drive schedule performance (Arena et al., 2013).

In closing, it is important to keep in mind that cost, schedule, and performance priorities should be balanced and set carefully based on each individual program's characteristics and, ultimately, the military needs that they are intended to satisfy. Given the challenges associated with accelerating programs (e.g., securing adequate funds and developing technology), the benefits of an aggressive program schedule without commensurate adjustments elsewhere are unclear. Even when there is an urgent operational need or adversaries have a critical technological edge, reason, discipline, and risk management must prevail. DoD's efforts to improve internal processes may help streamline the acquisition process and optimize schedule performance. However, prioritization among cost, schedule, and performance in individual programs must be logically determined and made clear before schedules can be set. This prioritization and its implications should be determined and thoroughly discussed by the acquisition and user communities to help avoid future problems in acquiring warfighter capabilities.

---

[1] "A framing assumption is any explicit or implicit assumption that is central in shaping cost, schedule, and/or performance expectations" (Arena et al., 2013, p. xix).

APPENDIX

A Case Study in Fulfilling an Urgent Operational Need: The MRAP Acquisition Program

This appendix presents a case study of a unique acquisition program that had extraordinary military, White House, congressional, and public support and financial resources to fulfill a critical wartime need on a very ambitious schedule. The Mine-Resistant Ambush-Protected Vehicle (MRAP) program is widely held as an exemplary "rapid acquisition" program in that it successfully fulfilled a specific UON for improved protection from improvised explosive devices (IEDs) for the U.S. Marine Corps and Army on an unusually short timeline. While the program is heralded as a rapid-acquisition success story, there were a number of unique conditions that may not be easily replicated. Examples of these conditions include an existing commercial variant; relatively low technical, design, and integration risk; significant internal and external senior leadership support; and a nearly unlimited budget. However, there are useful lessons that can be learned from the MRAP experience that can be applied to other programs. These transferrable lessons include

- using proven or mature technologies
- keeping most requirements reasonably stable
- prioritizing schedule over cost
- compromising on some technical performance requirements.

Early in the conflicts in Afghanistan and Iraq, hundreds of U.S. service members were killed or injured by IEDs each year. U.S. casualties rose to more than 1,000 per year by 2009. A need for better protection from IEDs was badly needed because the vehicles used in these regions at that time were not designed to sustain blasts from below and, thus, provided inadequate protection. Interest in providing soldiers with armored vehicles specifically designed to protect against this threat spread from the war zone back to the highest levels of leadership in the Pentagon and Congress. This high-level attention and support provided a platform for the MRAP program to move quickly in producing a more adequate mechanism for dealing with IED threats. In response to a validated U.S. Central Command JUON statement, the MRAP program office was created on November 1, 2006 (Howitz, 2008). In May 2007, former Defense Secretary Robert Gates established an MRAP task force with the mission to provide "as many of those vehicles to our Soldiers and Marines in the field as is possible in the next several months" (Howitz, 2008). The program was delivering a significant number of MRAPs by May 2008, less than two years after the program's official inception (GAO, 2009).

Several conditions enabled the quick production and fielding of MRAP vehicles:

- The program was designated as "DoD's highest priority acquisition" and was provided a DX rating by the Secretary of Defense in 2007 (GAO, 2009).[1] This special status afforded the program special contracting privileges, which helped it adhere to its ambitious schedule and motivated greater collaboration and cooperation among the acquisition, test, and user communities.
- Program officials developed a creative acquisition and contracting strategy, and they managed the program well.
- The capability gap that the MRAPs were meant to fulfill required minimal technology development, allowing the program to move swiftly to the production phase with multiple responsive and capable vendors (Howitz, 2008).
- Perhaps most importantly, the schedule was designated as the top priority for the program from the outset, which drove deci-

[1] DX is the code for the highest-priority defense programs in the Defense Priorities and Allocations System.

sionmaking and management behavior. This behavior may have differed under alternative (e.g., cost or reliability) considerations (Miller, 2010; Sullivan, 2009; Blakeman, Gibbs, and Jeyasingam, 2008).

Requirements: Defining and Maintaining Program Scope

The MRAP program was motivated by a specific, high-priority need in the U.S. Marine Corps in 2005, and, by 2006, it had been designated a JUON by U.S. Central Command. Such needs are defined in a JUON statement, which aims to provide validation and resourcing for a solution in an extremely short time period—usually within days or weeks (CJCSI 3170.01H, 2012). The specificity and high-priority status of the MRAP requirement ensured that the program was well protected throughout the procurement process.

The specific requirement that the MRAP program had to fulfill was for a vehicle that could better protect marines from IED blasts. The need for improved protection was met by increasing clearance above the ground, adding more armor, and designing a V-shaped hull that would be better able to deflect such blasts. These features would provide better protection than those offered by the high-mobility multipurpose wheeled vehicles that the Marine Corps was using at the time (Howitz, 2008). These specific design approaches did not require much technology development; in fact, similar vehicles had already been in use since the 1970s in South Africa and Rhodesia (now Zimbabwe; see Hodge, 2007). In addition, as discussed later, numerous eligible vendors were capable of producing the vehicles. The decisions to keep requirements to a minimum and to use proven technologies are widely credited with the program's ability to rapidly meet the needs of the Marine Corps and other services.[2] This strategy was successful despite instability in quantity orders and status. Over the course of three years, what began as a

[2] The Marine Corps originally planned to procure just 1,169 vehicles, but in 2007 the program became a joint program—serving all three services and U.S. Special Operations Command, with a joint requirement for over 16,000 vehicles. See Sullivan, 2009.

single-service, ACAT III program eventually became a joint, ACAT ID program (Blakeman, Gibbs, and Jeyasingam, 2008).

The MRAP's requirements remained relatively stable throughout the life of the program because of the fact that requirements changes could not be made without senior-level approval. Although there may have been good ideas that could have been incorporated later on, the MRAP program was able to avoid such "requirements creep" and keep the program moving forward (Miller, 2010).

Resource Allocation and Management

The higher priority a program has, the more likely it is to secure and keep the funding necessary to maintain its schedule. Because of its high-priority status, the MRAP program did not face many obstacles in terms of securing or defending funding and resources. Congressional support and a DX rating afforded the program nearly unlimited funding, including about $22.7 billion for procurement. These funds were provided primarily through supplemental appropriations and, at times, emergency appropriations and reprogramming actions (Miller, 2010). The MRAP program was essentially immune to the externally driven funding cuts or delays that other defense acquisition programs often face.

Tailoring the Acquisition Process and Setting Priorities

To meet its aggressive timeline, the MRAP program pursued a tailored acquisition plan, including approval to perform "simultaneous execution of all facets of the DoD acquisition framework" (Blakeman, Gibbs, and Jeyasingam, 2008). While the program was still required to provide all required acquisition documentation, many of the required documents were provided after the decisions they were supposed to support. The program did not have an APB prior to starting procurement at Milestone C in February 2007. The APB was not approved until June 16, 2008. Typically, an APB is required before procurement can begin. According to the program's December 2009 Selected Acquisition Report, the full-rate production (FRP) milestone was not required for the MRAP program "because the total program objective is being achieved via a series of Low Rate Initial Production (LRIP) orders reviewed and approved individually by the [Milestone Decision Authority]. The Joint MRAP has already acquired 22,880 of the 26,882 (Acquisition Objective) vehicle requirement negating the requirement for an FRP decision" (DoD, 2009). The capability production document (CPD) also lagged the decisions it was supposed to support. The CPD provides the operational requirements of the production system, including key performance parameters (DoDI 5000.02, 2008). MRAP production contracts were awarded and testing was under way before the CPD was approved by the Joint Requirements Oversight Council.

All this is not to say that there was insufficient oversight of the MRAP program. On the contrary, extensive decision meetings, reviews, and ad hoc oversight were conducted outside the regular process calendars. These examples, however, show how formal reports and processes can be delayed or adjusted based on operational need.

Concurrency

For the MRAP program, the use of proven technologies facilitated a high level of concurrency. Blakeman, Gibbs, and Jeyasingam (2008, p. 30) noted that "the MRAP program simultaneously conducted developmental testing, operational testing, production, integration, fielding, and disposal, while also refining requirements to account for an increasing Explosively Formed Penetrator (EFP) threat and greater need in the restrictive terrain of Afghanistan."

Contract Structures and Competition

Many contracting activities that normally occur in sequence were conducted in parallel to expedite the contracting process for the MRAP program. To begin production, a sole-source contract was awarded to a company with an active production line, while a request for proposal was released to industry. The joint program office received bids from ten companies. Because of the unusually high number of capable manufacturers, nine IDIQ contracts[3] were awarded for test vehicles. While the source selection criteria focused heavily on survivability, significant schedule incentives were specified. An incentive award of $100,000 was provided for each test vehicle delivered early (Blakeman, Gibbs, and Jeyasingam, 2008, citing an internal program document). Five companies were awarded large LRIP contracts prior to testing. A contracting mechanism established costs and prices with vendors up front, allowing quick remediation of contract modifications and helping to expedite schedules and minimize costs by avoiding lengthy negotiations. Without such an approach, every engineering change proposal would have required negotiation.

One of the biggest challenges to rapid procurement was quickly finding available government contracting officers who were skilled in executing such large, complex contracts. Also key was getting the Defense Contract Management Agency to work closely with all stakeholders to ensure delivery of the vehicles on the streamlined schedule. Conditional acceptance of vehicles with minor issues was allowed, which also sped delivery. Finally, a good relationship between vendors and the Space and Naval Warfare Systems Command helped the government with integration; this was important because the government maintained responsibility for integrating mission equipment with the vehicles.

Conclusion

In summary, the MRAP program's exceptional stakeholder support and resulting process flexibility are difficult to replicate—at least in a program that does not fulfill such an urgent need. Furthermore, the program did not suffer from technical risk, and key requirements

---

[3] See the Federal Acquisition Regulation, undated, Subpart 16.5, for information on indefinite-delivery contracts.

were not challenged. However, the MRAP program demonstrated that a creative acquisition strategy supported by adequate resourcing, cooperation between the acquisition and user communities, and effective program management can greatly expedite acquisition.

Bibliography

Anderson, Joshua, and Jeff Upton, "Unleashing the Predictive Power of the Integrated Master Schedule: The Planning and Scheduling Excellence Guide (PASEG)," *Defense AT&L*, January–February 2012, pp. 34–38.

Arena, Mark V., and John Birkler, *Determining When Competition Is a Reasonable Strategy for the Production Phase of Defense Acquisition*, Santa Monica, Calif.: RAND Corporation, OP-263-OSD, 2009. As of April 4, 2013: http://www.rand.org/pubs/occasional_papers/OP263.html

Arena, Mark V., John Birkler, John F. Schank, Jessie Riposo, and Clifford A. Grammich, *Monitoring the Progress of Shipbuilding Programmes: How Can the Defence Procurement Agency More Accurately Monitor Progress?* Santa Monica, Calif.: RAND Corporation, MG-235-MOD, 2005. As of June 12, 2013: http://www.rand.org/pubs/monographs/MG235.html

Arena, Mark V., Irv Blickstein, Abby Doll, Jeffrey A. Drezner, James G. Kallimani, Jennifer Kavanagh, Daniel F. McCaffrey, Megan McKernan, Charles Nemfakos, Rena Rudavsky, Jerry M. Sollinger, Daniel Tremblay, and Carolyn Wong, *Management Perspectives Pertaining to Root Cause Analyses of Nunn-McCurdy Breaches, Volume 4: Program Manager Tenure, Oversight of Acquisition Category II Programs, and Framing Assumptions*, Santa Monica, Calif.: RAND Corporation, MG-1171/4-OSD, 2013. As of November 3, 2013: http://www.rand.org/pubs/monographs/MG1171z4.html

Arena, Mark V., Irv Blickstein, Obaid Younossi, and Clifford A. Grammich, *Why Has the Cost of Navy Ships Risen? A Macroscopic Examination of the Trends in U.S. Naval Ship Costs Over the Past Several Decades*, Santa Monica, Calif.: RAND Corporation, MG-484-NAVY, 2006a. As of August 22, 2013: http://www.rand.org/pubs/monographs/MG484.html

Arena, Mark V., Robert S. Leonard, Sheila E. Murray, and Obaid Younossi, *Historical Cost Growth of Completed Weapon System Programs*, Santa Monica, Calif.: RAND Corporation, TR-343-AF, 2006b. As of June 12, 2013: http://www.rand.org/pubs/technical_reports/TR343.html

Arena, Mark V., Obaid Younossi, Lionel A. Galway, Bernard Fox, John C. Graser, Jerry M. Sollinger, Felicia Wu, and Carolyn Wong, *Impossible Certainty: Cost Risk Analysis for Air Force Systems*, Santa Monica, Calif.: RAND Corporation, MG-415-AF, 2006c. As of June 12, 2013:
http://www.rand.org/pubs/monographs/MG415.html

Birkler, John, John C. Graser, Mark V. Arena, Cynthia R. Cook, Gordon T. Lee, Mark A. Lorell, Giles K. Smith, Fred Timson, Obaid Younossi, and Jon Grossman, *Assessing Competitive Strategies for the Joint Strike Fighter: Opportunities and Options*, Santa Monica, Calif.: RAND Corporation, MR-1362-OSD/JSF, 2001. As of June 12, 2013:
http://www.rand.org/pubs/monograph_reports/MR1362.html

Blakeman, Seth T., Anthony R. Gibbs, and Jeyanthan Jeyasingam, *Study of the Mine Resistant Ambush Protected (MRAP) Vehicle Program as a Model for Rapid Defense Acquisitions*, Monterey, Calif.: Naval Postgraduate School, 2008.

Blickstein, Irv, Michael Boito, Jeffrey A. Drezner, James Dryden, Kenneth Horn, James G. Kallimani, Martin C. Libicki, Megan McKernan, Roger C. Molander, Charles Nemfakos, Chad J. R. Ohlandt, Caroline Reilly, Rena Rudavsky, Jerry M. Sollinger, Katharine Watkins Webb, and Carolyn Wong, *Root Cause Analyses of Nunn-McCurdy Breaches, Volume 1:* Zumwalt-*Class Destroyer, Joint Strike Fighter, Longbow Apache, and Wideband Global Satellite*, Santa Monica, Calif.: RAND Corporation, MG-1171/1-OSD, 2011. As of June 12, 2013:
http://www.rand.org/pubs/monographs/MG1171z1.html

Blickstein, Irv, Jeffrey A. Drezner, Martin C. Libicki, Brian McInnis, Megan McKernan, Charles Nemfakos, Jerry M. Sollinger, and Carolyn Wong, *Root Cause Analyses of Nunn-McCurdy Breaches, Volume 2: Excalibur Artillery Projectile and the Navy Enterprise Resource Planning Program, with an Approach to Analyzing Complexity and Risk*, Santa Monica, Calif.: RAND Corporation, MG-1171/2-OSD, 2012a. As of August 19, 2013:
http://www.rand.org/pubs/monographs/MG1171z2.html

Blickstein, Irv, Jeffrey A. Drezner, Brian McInnis, Megan McKernan, Charles Nemfakos, Jerry M. Sollinger, and Carolyn Wong, *Methodologies in Analyzing the Root Causes of Nunn-McCurdy Breaches*, Santa Monica, Calif.: RAND Corporation, TR-1248-OSD, 2012b. As of June 12, 2013:
http://www.rand.org/pubs/technical_reports/TR1248.html

Blickstein, Irv, Chelsea Kaihoi Duran, Daniel Gonzales, Jennifer Lamping Lewis, Charles Nemfakos, Jessie Riposo, Rena Rudavsky, Jerry M. Sollinger, Daniel Tremblay, and Erin York, *Root Cause Analyses of Nunn-McCurdy Breaches, Volume 3: Joint Tactical Radio System, P-8A Poseidon, and Global Hawk Modifications*, Santa Monica, Calif.: RAND Corporation, 2013, not available to the general public.

Blickstein, Irv, and Giles K. Smith, *A Preliminary Analysis of Advance Appropriations as a Budgeting Method for Navy Ship Procurements*, Santa Monica, Calif.: RAND Corporation, MR-1527-NAVY, 2002. As of June 12, 2013: http://www.rand.org/pubs/monograph_reports/MR1527.html

Bliss, Gary R., "Reducing Acquisition Cycle-Time in Technology Insertion," briefing, DoD Acquisition Insight Days, April 22, 2009.

———, *Performance Assessments and Root Cause Analyses: 2012 Annual Report*, Washington, D.C.: Office of the Under Secretary of Defense for Acquisition, Technology, and Logistics, March 2012a.

———, "Observations from AT&L/PARCA's Root Cause Analyses," briefing, 9th Annual Acquisition Research Symposium, May 17, 2012b.

———, *Report to Congress on Performance Assessments and Root Cause Analyses*, Washington, D.C.: Office of the Under Secretary of Defense for Acquisition, Technology, and Logistics, 2013.

Bodilly, Susan J., *Case Study of Risk Management in the USAF LANTIRN Program*, Santa Monica, Calif.: RAND Corporation, N-3617-AF, 1993. As of June 12, 2013:
http://www.rand.org/pubs/notes/N3617.html

Boehm, Barry, and Jo Ann Lane, "DoD Systems Engineering and Management Implications for Evolutionary Acquisition of Major Defense Systems: A DoD SERC Quick-Look Study and CSER 2010 Invited Presentation," *Systems Engineering Research Center (SERC)*, USC-CSSE-2010-500, March 17, 2010.

Bolten, Joseph G., Robert S. Leonard, Mark V. Arena, Obaid Younossi, and Jerry M. Sollinger, *Sources of Weapon System Cost Growth: Analysis of 35 Major Defense Acquisition Programs*, Santa Monica, Calif.: RAND Corporation, MG-670-AF, 2008. As of June 12, 2013:
http://www.rand.org/pubs/monographs/MG670.html

Bowsher, Charles A., Comptroller General of the United States, *Schedule Delays and Cost Overruns Plague DOD Automated Information Systems*, testimony before the Subcommittee on Legislation and National Security, Committee on Government Operations, U.S. House of Representatives, GAO/T-IMTEC-89-8, Washington, D.C.: U.S. General Accounting Office, May 18, 1989.

Brodfuehrer, Brian, "Cycle Time Reduction: A Total Systems Life Cycle View on Reducing Cycle Time," *Program Manager*, Vol. XXIX, No. 3, Ft. Belvoir, Va.: Defense Systems Management College Press, May–June 2000, pp. 22–27.

Carter, Ashton, Under Secretary of Defense for Acquisition, Technology, and Logistics, "Better Buying Power: Guidance for Obtaining Greater Efficiency and Productivity in Defense Spending," memorandum, September 14, 2010a.

———, "Implementation Directive for Better Buying Power—Obtaining Greater Efficiency and Productivity in Defense Spending," memorandum, November 3, 2010b.

Cashman, William M., *Why Schedules Slip: Actual Reasons for Schedule Problems Across Large Air Force System Development Efforts*, thesis, Wright-Patterson Air Force Base, Ohio: Air Force Institute of Technology, November 6, 1995.

Chairman of the Joint Chiefs of Staff Instruction 3170.01H, *Joint Capabilities Integration and Development System*, January 10, 2012. As of August 21, 2013: http://www.dtic.mil/cjcs_directives/cdata/unlimit/3170_01.pdf

Chittenden, James L., *Faster Is Better: Can the USAF Acquisition Process Be Saiv'd?* BibloScholar, November 13, 2012.

CJCSI—*See* Chairman of the Joint Chiefs of Staff Instruction.

Comptroller General of the United States, *Acquisition of Major Weapons Systems*, B-163058, Washington, D.C., March 18, 1971.

———, *The Importance of Testing and Evaluation in the Acquisition Process for Major Weapons Systems*, Washington, D.C., August 7, 1972.

———, *Relative Performance of Defense and Commercial Communications Satellite Programs*, Washington, D.C., LCD-79-108, August 10, 1979.

DAU—*See* Defense Acquisition University.

Decker, Gilbert F., Louis C. Wagner, William H. Forster, David M. Maddox, George T. Singley, and George G. Williams, *Army Strong: Equipped, Trained and Ready: Final Report of the 2010 Army Acquisition Review*, January 2011.

Defense Acquisition University, *Scheduling Guide for Program Managers*, Fort Belvoir, Va.: Defense Systems Management College Press, October 2001. As of December 17, 2013:
https://acc.dau.mil/adl/en-US/37441/file/8993/
Scheduling%20Guide%20for%20Program%20Managers.pdf

———, "Joint Rapid Acquisition Cell (JRAC)," last updated April 17, 2008. As of August 19, 2013:
https://acc.dau.mil/CommunityBrowser.aspx?id=189479&lang=en-US

———, *Defense Acquisition Guidebook*, Washington, D.C., last updated June 2013. As of August 22, 2013:
https://acc.dau.mil/dagfp/default.aspx

Defense Science Board, *Report of the Defense Science Board Task Force on Department of Defense Policies and Procedures for the Acquisition of Information Technology*, Washington, D.C.: Office of the Under Secretary of Defense for Acquisition, Technology, and Logistics, March 2009a. As of August 22, 2013: http://www.acq.osd.mil/dsb/reports/ADA498375.pdf

———, *Report of the Defense Science Board Task Force on the Fulfillment of Urgent Operational Needs*, Washington, D.C.: Office of the Under Secretary of Defense for Acquisition, Technology, and Logistics, July 2009b. As of June 13, 2013:
http://www.acq.osd.mil/dsb/reports/ADA503382.pdf

DoD—*See* U.S. Department of Defense.

DoDD—*See* U.S. Department of Defense Directive.

DoDI—*See* U.S. Department of Defense Instruction.

Drezner, Jeffrey A., "Competition and Innovation Under Complexity," in Guy Ben-Ari and Pierre A. Chao, eds., *Organizing for a Complex World: Developing Tomorrow's Defense and Net-Centric Systems*, Washington, D.C.: Center for Strategic and International Studies, 2009, pp. 31–49. As of June 12, 2013:
http://www.rand.org/pubs/reprints/RP1386

Drezner, Jeffrey A., Mark V. Arena, Megan McKernan, Robert Murphy, and Jessie Riposo, *Are Ships Different? Policies and Procedures for the Acquisition of Ship Programs*, Santa Monica, Calif.: RAND Corporation, MG-991-OSD/NAVY, 2011. As of June 12, 2013:
http://www.rand.org/pubs/monographs/MG991.html

Drezner, Jeffrey A., and Meilinda Huang, *On Prototyping: Lessons from RAND Research*, Santa Monica, Calif.: RAND Corporation, OP-267-OSD, 2009. As of November 4, 2013:
http://www.rand.org/pubs/occasional_papers/OP267.html

Drezner, Jeffrey A., Jeanne M. Jarvaise, Ron Hess, Daniel M. Norton, and Paul G. Hough, *An Analysis of Weapon System Cost Growth*, Santa Monica, Calif.: RAND Corporation, MR-291-AF, 1993. As of June 12, 2013:
http://www.rand.org/pubs/monograph_reports/MR291.html

Drezner, Jeffrey A., and Giles K. Smith, *An Analysis of Weapon System Acquisition Schedules*, Santa Monica, Calif.: RAND Corporation, R-3937-ACQ, 1990. As of June 12, 2013:
http://www.rand.org/pubs/reports/R3937.html

DSB—*See* Defense Science Board.

Farr, John V., William R. Johnson, and Robert P. Birmingham, "A Multitiered Approach to Army Acquisition," *Defense Acquisition Review Journal*, Vol. 12, No. 2, April–July 2005, pp. 235–246.

Fast, William R., "Improving Defense Acquisition Decision Making," *Defense Acquisition Review Journal*, Vol. 17, No. 2, April 2010, pp. 220–241.

Federal Acquisition Regulation, "Subpart 16.5—Indefinite-Delivery Contracts," undated. As of February 18, 2014:
http://www.acquisition.gov/far/current/html/Subpart%2016_5.html

Fox, Bernard, Michael Boito, John C. Graser, and Obaid Younossi, *Test and Evaluation Trends and Costs for Aircraft and Guided Weapons*, Santa Monica, Calif.: RAND Corporation, MG-109-AF, 2004. As of June 12, 2013: http://www.rand.org/pubs/monographs/MG109.html

Fox, J. Ronald, *Defense Acquisition Reform, 1960–2009: An Elusive Goal*, Washington, D.C.: Center of Military History, U.S. Army, 2011.

Gailey, Charles K., *Predictive Power for Program Success from Engineering and Manufacturing Development Performance Trends*, Fort Belvoir, Va.: Defense Acquisition University, November 2002.

GAO—*See* U.S. General Accounting Office (prior to July 7, 2004); U.S. Government Accountability Office (as of July 7, 2004).

Glennan, Thomas K., Susan J. Bodilly, Frank Camm, Kenneth R. Mayer, and Tim Webb, *Barriers to Managing Risk in Large Scale Weapons System Development Programs*, Santa Monica, Calif.: RAND Corporation, MR-248-AF, 1993. As of June 12, 2013:
http://www.rand.org/pubs/monograph_reports/MR248.html

Greenhouse, Bunnatine H., "Indefinite-Delivery/Indefinite-Quantity (IDIQ) Contracts," July 13, 2000.

Hanks, Christopher, Elliot Axelband, Shuna Lindsay, Mohammed Rehan Malik, and Brett Steele, *Reexamining Military Acquisition Reform: Are We There Yet?* Santa Monica, Calif.: RAND Corporation, MG-291-A, 2005. As of June 12, 2013: http://www.rand.org/pubs/monographs/MG291.html

Harmon, Bruce R., Lisa M. Ward, and Paul R. Palmer, *Assessing Acquisition Schedules for Tactical Aircraft*, Alexandria, Va.: Institute for Defense Analyses, P-2105, February 1989.

Held, Bruce J., *Improving the Department of Defense's Small Business Innovation Research Program*, testimony before the Subcommittee on Technology and Innovation, Science and Technology Committee, U.S. House of Representatives, April 26, 2007. As of June 17, 2013:
http://www.rand.org/pubs/testimonies/CT280.html

Held, Bruce J., Thomas R. Edison, Jr., Shari Lawrence Pfleeger, Philip S. Antón, and John Clancy, *Evaluation and Recommendations for Improvement of the Department of Defense Small Business Innovation Research (SBIR) Program*, Santa Monica, Calif.: RAND Corporation, DB-490-OSD, 2006. As of August 22, 2013:
http://www.rand.org/pubs/documented_briefings/DB490.html

Hodge, Nathan, "Making Way for MRAP: USMC Mine Resistant Ambush Protected Vehicle Update," *Jane's Defence Weekly*, September 20, 2007.

Howitz, Michael C., *The Mine Resistant Ambush Protected Vehicle: A Case Study*, Carlisle Barracks, Pa.: U.S. Army War College, March 2008.

Johnson, Collie J., "Pentagon Systems Acquisition Director Speaks to Graduates of APMC 99-1," *PM Magazine*, May–June 1999, pp. 8–11.

Kassing, David, R. William Thomas, Frank Camm, and Carolyn Wong, *How Funding Instability Affects Army Programs*, Santa Monica, Calif.: RAND Corporation, MG-447-A, 2007. As of February 20, 2013: http://www.rand.org/pubs/monographs/MG447.html

Kendall, Frank, Under Secretary of Defense for Acquisition, Technology, and Logistics, "Better Buying Power 2.0: Continuing the Pursuit for Greater Efficiency and Productivity in Defense Spending," memorandum, Washington, D.C.: U.S. Department of Defense, November 13, 2012a.

———, "Advance Questions for Frank Kendall, Nominee to Be Under Secretary of Defense for Acquisition, Technology, and Logistics," March 29, 2012b. As of October 25, 2013: http://www.acq.osd.mil/docs/Kendall%20APQs%20-%2029Mar2012.pdf

———, "Implementation Directive for Better Buying Power 2.0: Achieving Greater Efficiency and Productivity in Defense Spending," memorandum, April 24, 2013.

Kendall, Frank, David Van Buren, and David J. Venlet, "Tactical Air and Land Forces," Written Testimony for the House Armed Services Committee, Subcommittee on Tactical Air and Land Forces, U.S. House of Representatives, Combined Statement, Washington, D.C.: U.S. Department of Defense, March 20, 2012. As of October 25, 2013: http://www.acq.osd.mil/docs/HASC%20Subcommittee%20Hearing%20Testimony%20-%2020Mar2012.pdf

Lapham, Mary Ann, Ray Williams, Charles (Bud) Hammons, Daniel Burton, and Alfred Schenker, *Considerations for Using Agile in DoD Acquisition*, Pittsburgh, Pa.: Software Engineering Institute, Carnegie Mellon University, CMU/SEI-2010-TN-002, 2010. As of June 13, 2013: http://www.sei.cmu.edu/library/abstracts/reports/10tn002.cfm

Lorell, Mark A., and John C. Graser, *An Overview of Acquisition Reform Cost Savings Estimates*, Santa Monica, Calif.: RAND Corporation, MR-1329-AF, 2001. As of June 12, 2013: http://www.rand.org/pubs/monograph_reports/MR1329.html

Lorell, Mark A., Julia F. Lowell, and Obaid Younossi, *Evolutionary Acquisition: Implementation Challenges for Defense Space Programs*, Santa Monica, Calif.: RAND Corporation, MG-431-AF, 2006a. As of June 12, 2013: http://www.rand.org/pubs/monographs/MG431.html

———, *"Evolutionary Acquisition" Is a Promising Strategy, but Has Been Difficult to Implement*, Santa Monica, Calif.: RAND Corporation, RB-194-AF, 2006b. As of July 2, 2013: http://www.rand.org/pubs/research_briefs/RB194.html

Lush, Mona, Office of the Under Secretary of Defense for Acquisition, Technology, and Logistics, "Implementation of Weapon Systems Acquisition Reform Act (WSARA) of 2009 (Public Law 111-23, May 22, 2009)," briefing, October 22, 2009.

Mayer, Kenneth R., *The Development of the Advanced Medium Range Air-to-Air Missile: A Case Study of Risk and Reward in Weapon System Acquisition*, Santa Monica, Calif.: RAND Corporation, N-3620-AF, 1993. As of June 12, 2013: http://www.rand.org/pubs/notes/N3620.html

McCaffery, Jerry L., and Lawrence R. Jones, "Reform of Program Budgeting in the Department of Defense," *International Public Management Review*, Vol. 6, No. 2, 2005, pp. 141–176.

Miller, Thomas H., "Does MRAP Provide a Model for Acquisition Reform?" *Defense AT&L*, Vol. 39, No. 4, July–August 2010, pp. 16–20.

MITRE, "Systems Engineering Guide: Evolutionary Acquisition," web page, last updated May 8, 2012. As of August 19, 2013: http://www.mitre.org/publications/systems-engineering-guide/acquisition-systems-engineering/program-acquisition-strategy-formulation/evolutionary-acquisition

Monaco, James V., and Edward D. White III, "Investigating Schedule Slippage," *Defense Acquisition Review Journal*, April–July 2005, pp. 176–193.

Moore, Nancy Y., Amy G. Cox, Clifford A. Grammich, and Judith D. Mele, *Assessing the Impact of Requiring Justification and Approval Review for Sole Source 8(a) Native American Contracts in Excess of $20 Million*, Santa Monica, Calif.: RAND Corporation, TR-1011-OSD, 2012. As of June 12, 2013: http://www.rand.org/pubs/technical_reports/TR1011.html

Myers, Dominique, "Acquisition Reform—Inside the Silver Bullet: A Comparative Analysis—JDAM Versus F-22," *Acquisition Review Quarterly*, Fall 2002, pp. 313–322.

Nelson, Eric K., and Jay E. Trageser, *Analogy Selection Methodology Study*, Fairborn, Ohio: The Analytic Sciences Corporation, TR-5306-7-2, December 1987.

Neves, Sue, and Jack Strauss, "Survival Guide for Truly Schedule-Driven Development Programs," *Defense AT&L*, July–August 2008, pp. 21–23.

Office of the Inspector General, U.S. Department of Defense, *Audit of Major Defense Acquisition Programs Cycle Time*, Report No. D-2002-032, December 28, 2001.

Office of the Under Secretary of Defense for Acquisition, Technology, and Logistics, "Analyzing Profit or Fee," in *Contract Pricing Reference Guides, Volume 3: Cost Analysis*, Washington, D.C.: February 22, 2012. As of June 13, 2013: https://acc.dau.mil/CommunityBrowser.aspx?id=379516

———, *Performance of the Defense Acquisition System: 2013 Annual Report*, Washington, D.C., June 28, 2013.

OUSD(AT&L)—*See* Office of the Under Secretary of Defense for Acquisition, Technology, and Logistics.

Packard, David, Ernest C. Arbuckle, Robert H. Barrow, Nicholas F. Brady, Louis W. Cabot, Frank C. Carlucci, William P. Clark, Barber B. Conable, Jr., Paul F. Gorman, Carla A. Hills, James L. Holloway, III, William J. Perry, Charles J. Pilliod, Jr., Brent Scowcroft, Herbert Stein, R. James Woolsey, and Rhett B. Dawson, *A Quest for Excellence: Final Report to the President by the President's Blue Ribbon Commission on Defense Management*, June 30, 1986.

Parrish, Karen, "Defense Officials Preview 'Better Buying Power 2.0' Initiative," American Forces Press Service, November 13, 2012. As of January 24, 2013: http://www.defense.gov/News/NewsArticle.aspx?ID=118534

Pernin, Christopher G., Elliot Axelband, Jeffrey A. Drezner, Brian B. Dille, John Gordon IV, Bruce J. Held, K. Scott McMahon, Walter L. Perry, Christopher Rizzi, Akhil R. Shah, Peter A. Wilson, and Jerry M. Sollinger, *Lessons from the Army's Future Combat Systems Program*, Santa Monica, Calif.: RAND Corporation, MG-1206-A, 2012. As of June 12, 2013:
http://www.rand.org/pubs/monographs/MG1206.html

Peterson, Kyle, "A Wing and a Prayer: Outsourcing at Boeing," Reuters, January 20, 2011. As of August 22, 2013:
http://www.reuters.com/article/2011/01/20/us-boeing-dreamliner-idUSTRE70J2UX20110120

Portier, Bertrand, "SOA Terminology Overview, Part 1: Service, Architecture, Governance, and Business Terms," IBM website, May 24, 2007. As of February 25, 2014:
http://www.ibm.com/developerworks/library/ws-soa-term1/

Public Law 103-62, Government Performance and Results Act of 1993, January 5, 1993.

Public Law 111-23, Weapon Systems Acquisition Reform Act of 2009, May 22, 2009.

Reig, Raymond W, "A Decade of Success and Failures in the DoD Acquisition System: A Continuing Quality Journey," *Program Manager*, Vol. 24, No. 1, January–February 1995, pp. 27–29.

Riposo, Jessie, Guy Weichenberg, Chelsea Kaihoi, Bernard Fox, William Shelton, and Andreas Thorsen, *Improving Air Force Enterprise Resource Planning-Enabled Business Transformation*, Santa Monica, Calif.: RAND Corporation, RR-250-AF, 2013. As of January 27, 2014:
http://www.rand.org/pubs/research_reports/RR250.html

Scheduling Guide for Program Managers, Fort Belvoir, Va.: Defense Systems Management College Press, October 2001.

Schinasi, Katherine V., U.S. Government Accountability Office, *Defense Acquisitions: Better Weapon Program Outcomes Require Discipline, Accountability, and Fundamental Changes in the Acquisition Environment*, testimony before the Committee on Armed Services, U.S. Senate, Washington, D.C., GAO-08-782T, June 3, 2008.

Sullivan, Michael J., U.S. Government Accountability Office, *Defense Acquisitions: Rapid Acquisition of MRAP Vehicles*, Washington, D.C., GAO-10-155T, October 2009.

———, "Schedule-Driven Costs in Major Defense Programs," in *Proceedings from the Ninth Annual Research Symposium*, Monterey, Calif.: Naval Postgraduate School, April 30, 2012.

Thompson, Mark, "Indefinite-Delivery/Indefinite-Quantity," *Time*, February 26, 2013. As of June 13, 2013:
http://nation.time.com/2013/02/26/indefinite-deliveryindefinite-quantity

Tyson, Karen W., J. Richard Nelson, D. Calvin Gogerty, Bruce R. Hannon, and Alec W. Salerno, *Prototyping Defense Systems*, Alexandria, Va.: Institute for Defense Analyses, December 1991. As of June 13, 2013:
http://www.dtic.mil/dtic/tr/fulltext/u2/a252657.pdf

Tyson, Karen W., J. Richard Nelson, Neang I. Om, and Paul R. Palmer, *Acquiring Major Systems: Cost and Schedule Trends and Acquisition Initiative Effectiveness*, Alexandria, Va.: Institute for Defense Analyses, P-2201, March 1989.

U.S. Army Audit Agency, *Army Rapid Acquisition Processes: Tailored Acquisition*, May 9, 2011, not available to the general public.

U.S. Department of Defense, *Government Performance and Results Act: Department of Defense FY 2000 Performance Report*, Washington, D.C., March 2001. As of June 13, 2013:
http://www.finance.hq.navy.mil/fmb/gpra/GPRA_FY2000-Report.pdf

———, *U.S. Department of Defense Extension to A Guide to the Project Management Body of Knowledge (PMBOK® Guide)*, 1st ed., version 1.0, Fort Belvoir, Va.: Defense Acquisition University Press, June 2003.

———, *Joint Mine Resistant Ambush Protected Vehicle (JOINT MRAP)*, Selected Acquisition Report, December 31, 2009.

———, *Sustaining U.S. Global Leadership: Priorities for 21st Century Defense*, Washington, D.C., January 2012a.

———, "Performance Improvement," in *FY 2013 Budget Proposal: Overview*, Washington, D.C., February 13, 2012b. As of August 22, 2013:
http://comptroller.defense.gov/defbudget/fy2013/FY2013_Budget_Request_Overview_Book.pdf

U.S. Code, Title 10, Section 2366b, Major Defense Acquisition Programs: Certification Required Before Milestone B Approval.

U.S. Department of Defense Directive 5000.01, *The Defense Acquisition System*, May 12, 2003.

U.S. Department of Defense Directive 7045.14, *The Planning, Programming, Budgeting, and Execution (PPBE) Process*, January 25, 2013.

U.S. Department of Defense Instruction 5000.02, *Operation of the Defense Acquisition System,* December 8, 2008.

U.S. Department of Defense Instruction 5000.02, Interim, *Operation of the Defense Acquisition System,* November 25, 2013.

U.S. General Accounting Office, *Impediments to Reducing the Costs of Weapon Systems*, Washington, D.C., PSAD-80-6, November 1979.

———, *Sergeant York: Concerns About the Army's Accelerated Acquisition Strategy*, Washington, D.C., GAO/NSIAD-86-89, May 1986a.

———, *DoD's Defense Acquisition Improvement Program: A Status Report*, Washington, D.C., NSIAD-86-148, July 1986b.

———, *DoD's Acquisition Improvement Program: Program Managers' Views*, Washington, D.C., GAO/NSIAD-86-193FS, September 1986c.

———, *Acquisition: Status of the Defense Acquisition Improvement Program's 33 Initiatives*, Washington, D.C., GAO/NSIAD-86-178BR, September 1986d.

———, *Missile Procurement: AMRAAM Cost Growth and Schedule Delays*, Washington, D.C., NSIAD-87-78, March 1987.

———, *Major Acquisitions: Summary of Recurring Problems and Systemic Issues, 1960–1987*, Washington, D.C., GAO/NSIAD-88-135BR, September 1988.

———, *Weapons Testing: DoD Needs to Plan and Conduct More Timely Operational Tests and Evaluation*, Washington, D.C., GAO/NSIAD-90-107, May 1990.

———, *Tactical Missile Acquisitions: Understated Technical Risks Leading to Cost and Schedule Overruns*, Washington, D.C., GAO/NSIAD-91-280, September 1991.

———, *Best Practices: Better Management of Technology Development Can Improve Weapon System Outcomes*, Washington, D.C., GAO/NSIAD-99-162, July 1999.

U.S. Government Accountability Office, *Defense Acquisitions: Improvements Needed in Space Systems Acquisition Management Policy*, Washington, D.C., GAO-03-1073, September 2003.

———, *Defense Acquisitions: Major Weapon Systems Continue to Experience Cost and Schedule Problems under DoD's Revised Policy*, Washington, D.C., GAO-06-368, April 2006.

———, *Defense Acquisitions: Assessments of Selected Major Weapon Programs*, Washington, D.C., GAO-07-406, March 2007.

———, *Defense Acquisitions: Assessments of Selected Weapon Programs*, Washington, D.C., GAO-09-326SP, March 2009.

———, *Defense Acquisitions: Strong Leadership Is Key to Planning and Executing Stable Weapon Programs*, Washington, D.C., GAO-10-522, May 2010.

———, *Defense Acquisitions: Assessments of Selected Weapon Programs*, Washington, D.C., GAO-11-233SP, March 2011a.

———, *Actions Needed to Address Systems Engineering and Developmental Testing Challenges*, Washington, D.C., GAO-11-806, September 2011b.

———, *Defense Acquisitions: Assessments of Selected Major Weapon Programs*, Washington, D.C., GAO-12-400SP, March 2012a.

———, *Missile Defense: Opportunity Exists to Strengthen Acquisitions by Reducing Concurrency*, Washington, D.C., GAO-12-486, April 2012b.

———, *Urgent Warfighter Needs: Opportunities Exist to Expedite Development and Fielding of Joint Capabilities*, Washington, D.C., GAO-12-385, April 2012c.

———, *Best Practices for Project Schedules—Exposure Draft*, Washington, D.C., GAO-12-120G, May 2012d.

———, *KC-46 Tanker Aircraft: Program Generally Stable but Improvements in Managing Schedule Are Needed*, Washington, D.C., GAO-13-258, February 2013.

Vanden Brook, Tom, "Gates: MRAPs Save 'Thousands' of Troop Lives," *USA Today*, June 27, 2011.

Ward, Dan, and Chris Quaid, "It's About Time," *Defense AT&L*, January–February 2006, pp. 14–17.

Yakyma, Alex, and Dean Leffingwell, "The Principles of Agile Architecture," web page, last updated January 2, 2013. As of August 22, 2013:
http://scaledagileframework.com/the-principles-of-agile-architecture

Younossi, Obaid, Mark V. Arena, Kevin Brancato, John C. Graser, Benjamin W. Goldsmith, Mark A. Lorell, Fred Timson, and Jerry M. Sollinger, *F-22A Multiyear Procurement Program: An Assessment of Cost Savings*, Santa Monica, Calif.: RAND Corporation, MG-664-OSD, 2007. As of June 12, 2013:
http://www.rand.org/pubs/monographs/MG664.html

Younossi, Obaid, Mark A. Lorell, Kevin Brancato, Cynthia R. Cook, Mel Eisman, Bernard Fox, John C. Graser, Yool Kim, Robert S. Leonard, Shari Lawrence Pfleeger, and Jerry M. Sollinger, *Improving the Cost Estimation of Space Systems: Past Lessons and Future Recommendations*, Santa Monica, Calif.: RAND Corporation, MG-690-AF, 2008. As of June 12, 2013:
http://www.rand.org/pubs/monographs/MG690.html

Younossi, Obaid, David E. Stem, Mark A. Lorell, and Frances M. Lussier, *Lessons Learned from the F/A-22 and F/A-18 E/F Development Programs*, Santa Monica, Calif.: RAND Corporation, MG-276-AF, 2005. As of June 12, 2013:
http://www.rand.org/pubs/monographs/MG276.html